# Game-Changer

# Game-Changer

## GAME THEORY AND THE ART OF
## TRANSFORMING STRATEGIC SITUATIONS

## David McAdams

W. W. Norton & Company

*New York • London*

Excerpt from #54 [12 I.] from *The Way of Life According to Lao Tzu*
*Edited by Witter Bynner.* Copyright 1944 by Witter Bynner,
renewed © 1972 by Dorothy Chauvenet and Paul Horgan.
Reprinted by permission of HarperCollins Publishers.

"The Lockhorns' Night Out" on p. 95 is
© LOCKHORNS © W.HOEST ENTERPRISES, INC.
Distributed by King Features Syndicate.

For information about permission to reproduce selections from this book,
write to Permissions, W. W. Norton & Company, Inc.,
500 Fifth Avenue, New York, NY 10110

For information about special discounts for bulk purchases, please contact
W. W. Norton Special Sales at specialsales@wwnorton.com or 800-233-4830

Manufacturing by Courier Westford
Book design by Chris Welch
Production manager: Julia Druskin

The Library of Congress has catalogued the hardcover edition as follows:

McAdams, David.
Game-changer : game theory and the art of transforming strategic situations / David
McAdams. — First edition.
pages cm
Includes bibliographical references and index.
ISBN 978-0-393-23967-6 (hardcover)
1. Game theory. 2. Strategic planning. 3. Strategy. I. Title.
HB144.M327 2014
658.4'033—dc23

2013039711

ISBN 978-0-393-34989-4 pbk.

W. W. Norton & Company, Inc.
500 Fifth Avenue, New York, N.Y. 10110
www.wwnorton.com

W. W. Norton & Company Ltd.
Castle House, 75/76 Wells Street, London W1T 3QT

1 2 3 4 5 6 7 8 9 0

*For my children*
*Never forget how much I love you*

# CONTENTS

# Game-Changer

# PROLOGUE

The game-theory approach to business ... has resulted in
[many] strategic initiatives, from joint ventures to mergers
to new-business development, that would have been
unheard of in a traditional planning environment.

—*Raymond W. Smith, chairman of Bell Atlantic*

In a 1996 *Fortune* magazine article titled "Business as War Game:
A Report from the Battlefront," Raymond Smith wrote that "the
game-theory approach to business" was the secret to Bell Atlantic's success under his leadership as chairman during the telecom-
munications shake-up of the 1990s.[1] "Games" are strategic situations
and "game theory" is the art and science of strategy, but the game-
theory approach to business is much more than just being smart and
savvy about strategy. Indeed, as Smith explained, the game-theory
approach to business requires "a different kind of corporate manager:
flexible, intellectually rigorous, and highly tolerant of ambiguity" and
"a special kind of company [that] nurtures a climate of open, frank,
and relentlessly objective discussion so that all the variables are scru-
tinized honestly and without political repercussions."

My goal in this book is to introduce you to the game-theory
approach to life, in all its aspects, including business. At the heart of
the game-theory mind-set is the recognition that *the game can always
be changed.* This book will show you how to change the game, so that
you can then enjoy a consistent strategic advantage over your competitors. The approach here is unusual, in two ways. First, although

there are many different types of games worth knowing about, I will return time and time again to just one, the game known as the Prisoners' Dilemma. The Prisoners' Dilemma is important, with many wide-ranging applications, but the reason I focus on it is that game theory offers so many different ways to escape the "dilemma" in this game. As such, the Prisoners' Dilemma provides a natural showcase of the power and versatility of game theory in practice.

In Part One ("The Game-Changer's Toolkit"), we will explore six ways to change games (commitment, regulation, cartelization, retaliation, trust, and relationships), five of which allow one to escape the Prisoners' Dilemma, and three other key game-theory ideas (the timing of moves, strategic evolution, and equilibrium). Each game-changing approach will have its own chapter, while each of the other key ideas is the subject of a separate "Game-Theory Focus" section. My hope is that you will finish Part One with a real sense of mastery over the Prisoners' Dilemma, and with a toolkit of game-theory ideas that can be applied broadly to many other sorts of strategic situations.

The second unusual feature of this book is that I put my money where my mouth is and apply the game-theory approach that I preach to real strategic problems. Part Two ("The Game-Changer Files") presents six tales of pressing strategic problems, varying in urgency and importance from how to keep prices low on the Internet (File 1) and how to build trust on eBay (File 5) to, I kid you not, how to save mankind from looming and seemingly unstoppable dread disease (File 6). In each case, I use the game-theory approach to identify the strategic crux of the problem, and then leverage that "game-awareness" to brainstorm ideas on how to change the game to solve or at least mitigate the underlying problem.

So, get ready for a ride. You'll emerge a deeper strategic thinker, armed and ready to gain a strategic advantage in all the games you play, in business and in life.

# INTRODUCTION

The wise win before they fight, while the ignorant
fight to win.

—*Zhuge Liang, regent of the Shu kingdom, lived AD 181–234*

"The wise win before they fight." So wrote Zhuge Liang, the great statesman, scholar, and military commander of China's Three Kingdoms period.[1] That may sound like an empty platitude, but it captures an essential truth. The wise win before they fight by recognizing all the games that *could* be played, steering the strategic environment in their favor, and then fighting with confidence in their ultimate victory. By contrast, the ignorant just play the game that lies before them, their victory or defeat largely out of their control, a matter of luck and fortune.

I refer to the wisdom of Zhuge Liang as game-awareness, the ability to see the strategic world around you with open eyes. Game-awareness helps protect you from the many dangers of not knowing what games you are really playing. Moreover, once you are truly aware of the games in your life, you can take steps to change them to your strategic advantage. That's why, in addition to cultivating your game-awareness, my focus throughout this book is on how the lessons of game theory inform the art of changing games. Mastering this art will allow you to recognize and seize strategic opportunities that others do not see, giving you a significant advantage over your peers.

Over the past forty years, the science of game theory has risen from a fairly obscure branch of applied mathematics to the engine driving many of the most important intellectual advances in the social sciences. In the classroom, game theory is now a mainstay in a wide variety of fields, from economics and political science to business strategy, and is making inroads in disciplines such as law, corporate finance, managerial accounting, and social entrepreneurship, even biology and epidemiology.

Even if you've never heard of game theory, its lingo and concepts are in the air you breathe. What does America's planned troop withdrawal "signal" to the Afghan Taliban? Will a Greek debt default lead to "contagion" and financial crises elsewhere? Did Sprint's early investment in 4G WiMAX technology give them a "first-mover advantage"? These game-theoretic questions were all in the news in recent years.

## Game Theory in Business

In 2005, *Fast Company* magazine made a splash with an article claiming that no one uses game theory in business.[2] In their reporting, however, *Fast Company* doesn't appear to have spoken with any actual business leaders. Those people tell a different story: how game theory can and does give them and their businesses a strategic advantage.

First, game theory helps businesses plot tactics. The most obvious games in business are those played at the tactical level—how to set prices, how to launch a new product, and so on. Management consultants the world over use game theory when formulating tactical strategic advice on how to win such games.[3]

US military planners long ago learned the value of game theory for tactics. Before any major mission, they routinely play "war games," in which one group of officers is tasked with playing the enemy and achieving the enemy's objectives. War gaming is essential, as it exposes weaknesses in one's initial strategy and leads to a more robust final plan. On the other hand, a McKinsey global survey of over 1,800 business leaders found that about half don't even consider more than one of their own options when making important business decisions, much less how the competition might respond.[4] Of course, that just gives your firm a leg up if you can deploy game theory in a more meaningful way.

Second, game theory provides actionable insights. We are surrounded by games whose outcomes affect us, including many over which we have little control. Game theory provides conceptual insights that allow one to understand and predict, before others, what is likely to happen in such games. For example, according to Tom Copeland, chairman emeritus of corporate finance at Monitor, a leading strategy consulting firm: "Game theory can explain why oligopolies tend to be unprofitable, the cycle of overcapacity and overbuilding, and the tendency to execute real options earlier than optimal."[5]

Finally and most importantly, game theory can transform the culture of an organization. Firms are not simply players in games. They are also the milieu in which many games are played: among divisions, between workers and managers, between ownership and management, among stockholders and bondholders, and so on. Game theory realizes its greatest business potential when leaders of a firm create the culture and organizational structures needed for everyone to thrive together.

Raymond Smith's key insight, as stated at the start of the prologue, was that the process of planning business strategy is itself a game played within the firm. This game can often be dysfunctional and unproductive, as employees fear to openly question the status quo and managers defend the parochial interests of their divisions. To change this game for the better, it's essential to attract and/or cultivate a different sort of player ("flexible, intellectually rigorous, and highly tolerant of ambiguity") and to motivate everyone to contribute meaningfully to the planning process (by creating "a climate of open, frank, and relentlessly objective discussion . . . without political repercussions"). That's true, but planning is just the tip of the iceberg.

A game-aware management team can transform everything from how employees are motivated to how buyer and supplier relationships are nurtured, and much more.

Perhaps the best example of what game-awareness can accomplish in business is provided by Alfred Sloan, the legendary leader of General Motors (GM). Sloan is the paragon of game-awareness at the highest levels of modern management. As his brilliant autobiography *My Years with General Motors* (1963) makes abundantly clear, Sloan's ability to deeply understand the games of the automobile market profoundly transformed not only GM but the entire industry. For instance, Sloan's appreciation of the importance to consumers of fashion and aspiration led GM to introduce the annual model (a new design each year) and to encourage trade-ins of used cars for new. Similarly, Sloan's understanding of dealers' incentives and strategic (un)sophistication led GM to be the first manufacturer to offer to buy back unsold inventory, as well as to pioneer an integrated accounting system. Most importantly, Sloan recognized how each of his divisional managers had a competitive incentive to advance only his own division's interests. Transforming this game among his subordinates led Sloan to invent a new organizational form for the modern firm, as a confederacy of divisions, with a profound and enduring impact on American business.

## Game Theory for Strategic Advantage

Most people who have learned a little game theory can't imagine how they might use it in real life. Real-world strategic interactions are never as simple as the examples typically given in books or classes on game theory. It's often even unclear what game is really being played. In this book, we won't run from such complications or pretend they don't exist. Instead, we will embrace complexity and ambiguity as creating additional opportunities and avenues by which to change games to our advantage.

For instance, consider the notion that players are "rational." Rationality requires (i) a coherent view of the world and what one wants in life and (ii) consistent pursuit of that self-interest. But who among us can pass such a stringent test? Who among us knows what we really want—all the time, in every situation—and never succumbs to temptation or self-destruction? All in all, it's fairly obvious that no one is truly rational. Fortunately, game theory does not require rationality.[*] Indeed, game theory is perfectly suited to provide guidance on how to strategize in settings with potentially irrational players—including whether to act crazy yourself.

Several years ago, while a young professor at the MIT Sloan School of Management, I created a new type of business school course based on this deeper and more applicable vision of game theory. The course attracted just thirty students in its first year, 2004, as few were willing to take a chance on an untried class taught by a little-known professor. But those first thirty students experienced something unexpected. They emerged with eyes wide open to the world of games around them, ready to transform those games to their strategic advantage, and eager to spread the word to friends and colleagues. Sixty students enrolled in 2005, then 120 in 2006, after which "Game Theory for Strategic Advantage" became one of the most popular courses in the school. Student comments in spring 2008 included: "the best class at MIT Sloan," "fun, challenging, and useful," "incredibly effective," and "we will apply this in real life."

The best part of this course is a final project in which student teams (i) identify someone (real or imagined) who faces a strategic challenge of vital importance, and then (ii) provide wise advice in a persuasive, jargon-free memo. These projects run the gamut, including:

---

[*] In fact, game theory *predicts* irrational-seeming behavior in some contexts. In the financial world, for instance, "asset bubbles" occur, in which an asset persistently trades at a price above its inherent value. Recently, economists have shown how such bubbles can arise from a game played among investors and endure even after all investors know that the asset is overpriced. See, e.g., Dilip Abreu and Markus Brunnermeier, "Bubbles and Crashes," *Econometrica*, 2003.

- *business strategy*: e.g., the future of Google Wallet, the auto industry's response to TrueCar.com
- *public policy*: e.g., how best to direct resources to encourage New Orleanians to return after Hurricane Katrina
- *foreign policy*: e.g., how to tame the scourge of Somali piracy
- *sports*: e.g., how to spice up the NBA Slam Dunk Contest
- *home life*: e.g., how to get a toddler to sleep in her own bed
- *historical fiction*: e.g., how Pontius Pilate should deal with the troublesome case of Jesus the Nazarene
- *just plain silly*: e.g., how Elaine Benes can procure a fabulous Nicole Miller dress in episode 129 of the TV show *Seinfeld*

These student projects, as much as anything else, have converted me fully to the view that game theory—when used properly, with wisdom and humility—can be a powerful and positive transformative force. My goal in writing this book is nothing less than to spread this good news, to convert you too into a game-theory disciple, and to equip and empower you to employ game theory to maximum positive effect—not just to win the games you play, but to change those games and the strategic ecosystems in which they reside, to transform your life and our lives together for the better.

*When used with wisdom and humility, game theory can be a powerful and positive transformative force.*

Of course, we face many intractable problems that game theory alone cannot solve: close to home, in our families and workplaces; on the national stage, in our politics and public policy; and even as a species, striving to survive against disease and hateful ideology. However, even in these cases, clear-eyed strategic thinking can help identify key factors that cause or contribute to these problems. A proper application of game theory can then point the way to practical solutions, while also highlighting (before it's too late) unintended consequences that could make the cure worse than the disease.

That said, game theory is one of those tools that can cause trouble when used improperly. The process of modeling a game has a tendency to lull one into uncritical acceptance of the assumptions implicit in that model, false confidence in the predictions and recommendations that it generates, and intellectual blindness to changing conditions. Combating this modeler's malaise requires discipline, energy, and the rigorously flexible mind-set of a game theorist. Without such a mind-set and the game-awareness that it provides, game theory is worse than useless, even dangerous, to those who wield it.

# The Danger of Mathematical Theories

A little knowledge is a dangerous thing. So is a lot.

—*Albert Einstein*

According to a 2012 Gallup poll, 45 percent of Americans have a gun in their home.[6] Guns offer protection, but they also create new dangers. Fortunately, these dangers can be mitigated by training (e.g., learning to shoot safely and accurately) and by adopting best practices (e.g., storing your weapon out of children's reach). Mathematical theories are different, often only becoming truly dangerous in the hands of those with the *most* training and expertise.

## Example: Newton's Folly

I can calculate the motions of heavenly bodies, but not the madness of people.

—*Isaac Newton, 1720*

Sir Isaac Newton was the genius of his age. The inventor of Newtonian physics and co-inventor of calculus, Newton understandably believed that he could leverage his analytical skills to make money in stocks. After all, Newton knew more about the laws of motion than anyone else alive in his time. Certainly he could apply that knowledge to outperform the average broker or blacksmith speculator. And 1720 was an excellent year to make a *lot* of money trading in Britain's fledgling stock market. Prices were extremely volatile, and someone who could predict future price movements could make a killing.

Stock in the South Sea Company especially caught Newton's eye. Founded in 1711, the South Sea Company was granted a monopoly to trade in Spain's South American colonies.[7] Nothing excited early eighteenth-century investors more than the prospect of untold riches from trade with the New World. In 1720, this fervor led South Sea Company stock to rise tenfold, from £100 a share in January to nearly £1,000 a share in July, before it fell back down again to about £100 a share in December. Countless fortunes were made in the "South Sea Bubble" by those who rode the wave up and got off before the crash.[8] But for every big winner, there was an equally big loser.

Newton was one of the biggest losers, down £10,000 at a time when £200 was a comfortable annual income for a middle-class family. Newton complained in his diary that he was unable to fathom the "madness of people," as if his losses were their fault for not behaving as Newton had predicted. In fact, Newton had no one to blame but himself for his overconfidence in his own analysis.

Newton's laws of gravity and momentum apply to inanimate objects like planets and other heavenly bodies only because inanimate objects lack the *will* to pursue their own objectives. When NASA sends a new probe up to Mars, there are many complications and variables to consider, but there's one thing NASA doesn't need to worry about—that Mars will see them coming and get out of the way. But that's exactly what happens in games, including the stock market,[9] from Newton's time to our own.

## Example: Options and the Black–Scholes Formula

> It ain't what you don't know that gets you into trouble. It's
> what you know for sure that just ain't so.
>
> —*Mark Twain*

In 1973, Fischer Black, Myron Scholes, and Merton Miller published a pair of academic papers developing the theory of how to price options, an esoteric sort of financial contract that was rarely traded at the time. These papers transformed finance and would earn Scholes a Nobel Prize in Economics in 1997.[10] The crowning glory of this work is the Black–Scholes formula, which allows traders to identify when options are "mispriced" according to the theory. A new breed of "risk arbitrageurs" was born, all of them trading options on the basis of the Black–Scholes formula and making money hand over fist, at least for a while.

But then, in 1998, it all came crashing down with the fall of the hedge fund Long-Term Capital Management (LTCM) and the subsequent crisis in the financial markets. You see, there was one little problem. Just about every sophisticated, deep-pocketed financial investor was betting on the basis of the Black–Scholes formula, often with huge, dramatically leveraged bets that could themselves move markets. When one of those bets went bad in mid-1998 and everyone needed to sell to satisfy creditors, no one was available on the other side of the market to buy. This created a so-called "liquidity crisis" that not only destroyed LTCM but nearly brought down the entire market.*

Ironically, the Black–Scholes formula lost its accuracy and validity—thereby becoming dangerous knowledge—only when it became well known and widely adopted. As Nobelist Merton Miller himself noted after the fact: "The question . . . is whether the LTCM disaster was merely a unique and isolated event, a bad drawing from Nature's urn

---

*This is not hyperbole. William J. McDonough, president of the New York Federal Reserve, was quoted as saying that, absent intervention, "Markets would . . . probably cease to function" due to the LTCM crisis.

[i.e., just bad luck]; or whether such disasters are the inevitable consequence of the Black–Scholes formula itself and the illusion it may give
that all market participants can hedge away all their risk at the same
time."[11] More game-awareness might have allowed traders to avert
this crisis, by helping them to realize how their investment decisions
were strategically intertwined.

Isaac Newton and the architects of LTCM were brilliant and creative
mathematicians. How could they have failed to appreciate the limitations of their own analyses and the risks inherent in their investment
strategies? Part of the problem may have been their reliance on mathematics itself. Mathematics is built on logic and proof, often creating
the perception that mathematical arguments have authority over
intuition and even empirical observation. However, the "proof" that
mathematics offers is conditional on the assumptions that one brings
to the table. It's therefore essential that anyone who uses mathematics to make real-world decisions complements analysis with awareness of how the world really works, including what games are really
being played.

A deeper issue is how mathematics can change the way that we
view a situation. In one famous study,[12] college students were asked
to put themselves in a manager's shoes and decide how many workers
to fire during a hypothetical business slump. When the problem was
presented mathematically, with profits described by a formula rather
than shown in a table, students fired far more workers than otherwise. Even philosophy majors transformed into heartless suits once
their firing decision was framed in terms of a formula.

Presenting the problem mathematically caused students to think
about layoffs differently, focusing more on the bottom line than on
the people involved. Some business leaders might say that's a good
thing, that emotions and fellow feeling have no place in business.
But that's plainly wrong. Long-term profits depend on having a
motivated workforce, a loyal customer base, and a trusted supplier

network, none of which can be properly cultivated if one just focuses on next quarter's number.

Successful business leaders must invest in relationships, but how? Will a salary increase motivate employees to work smarter? Can cut-rate prices buy customer loyalty? We know that's not the whole story either, since the most energized and devoted workers often demand the lowest wage (e.g., volunteers for a charity) and the most loyal customers often pay the highest price (e.g., Mac and iPhone lovers). Charities and businesses like Apple have learned how to stoke and harness the passions of their workers and customers, to achieve lower labor costs and greater profit.

The same principle applies to all businesses, even boring ones. Stronger relationships translate into greater profits, but nurturing such relationships requires game-awareness to understand others' true motivations. More than that, game-awareness allows us to avoid the pitfalls and needless blunders that can come when we fail to anticipate the hidden players, hidden options, and hidden connections in games. In business, game-awareness failures can cost millions and embarrass a firm, while in war, game-unawareness can cost thousands of lives and even shape the destiny of nations.

Clearly, we need game-awareness in our boardrooms and among our military brass. More broadly, we need game-awareness in all avenues of life where a nexus of strategic interactions creates the possibility for dramatic failure or fantastic success. Perhaps most of all, we need more game-awareness in our schools and in our homes, so we can strengthen our families and prepare our children for a future that will be rich in strategic opportunities but rife with strategic dangers—a future where Game-Changers will thrive best.

# Part One

# THE GAME-CHANGER'S TOOLKIT

Successful business strategy is about actively shaping the
game you play, not just playing the game you find.

*—Adam Brandenburger and Barry Nalebuff,*
Co-opetition *(1997)*

The greatest power of game theory is to build your awareness of what games are being played and to illuminate ways in which those games can be changed for the better. The goal of this book is to introduce you to game theory at this deeper level, to set you on the path to becoming a Game-Changer in your own right, so that you can win *before* you fight, in business and in life, by actively shaping the games you play.

Leaders of all stripes—lawmakers and policymakers in government, CEOs and managers in business, mavens and trendsetters in society, professors and administrators in academia—face countless decisions that shape the games played by themselves and others. The lessons of this book can provide leaders with a powerful set of tools to leverage their influence and gain an even greater strategic advantage, to shape the world even more to their liking. But you don't need authority or influence or fame to use this book. Everyone has daily opportunities to shape the games that we play, making the lessons of game theory indispensable for all.

Part One develops the Game-Changer's Toolkit, six distinct approaches to change games to your strategic advantage:

1. Commit (chapter 1)
2. Invite regulation (chapter 2)
3. Merge or "collude" (chapter 3)
4. Enable retaliation (chapter 4)
5. Build trust (chapter 5)
6. Leverage relationships (chapter 6)

Part Two will then put these tools to work. (For those interested, updates and other additional material will be available at the website McAdamsGameChanger.com.)

CHAPTER 1

# Commit

Oh my warriors, whither would you flee? Behind you is the
sea, before you, the enemy. You have left now only the hope
of your courage and your constancy.

*—Tariq ibn Ziyad, general of the Umayyad caliphate, after*
*burning the fleet that carried his troops to Spain in AD 711*

I
n AD 711, Tariq ibn Ziyad led the Muslim army that would con-
quer the Iberian Peninsula. After crossing the Strait of Gibraltar
(which is named after him),[1] Tariq famously burned his fleet, just
before facing—and routing—the vastly more numerous forces of King
Roderick of the Visigoths. In fact, Tariq probably did not burn his
ships.[2] His men were recent converts to Islam, fierce Berber warriors
eager for a fight, so there was little need to steel their resolve. Furthermore, Tariq had already sown the seeds of near-certain victory,
by cultivating secret allies within Roderick's ranks who would turn
on their king at a decisive moment during the battle.

Even though Tariq the man probably didn't burn his fleet, Tariq the
legend did, and Spaniards repeated the tale for centuries. Certainly,
that legend was well known to the Spanish conquistador Hernán
Cortés when, in 1519, he took a page from Tariq's book and scuttled
all but one of his ships near Veracruz, Mexico, just before facing—and
routing—the vastly more numerous forces of King Moctezuma of the
Aztecs. Indeed, the parallels between Tariq and Cortés are so complete that it's hard to imagine Cortés did not make the connection.

Both men rose from humble origins (Tariq started as a slave, Cortés came from a family of "lesser nobility") to positions of great authority based on merit alone, during long campaigns of imperial conquest. Both led expeditions that wound up conquering a new land (Spain for Tariq, Mexico for Cortés), against a vastly more numerous native force (Visigoths for Tariq, Aztecs for Cortés). Both succeeded by recognizing and exploiting preexisting divisions among those natives. But neither had been given a mandate for their conquest. Indeed, both set forth in direct defiance of jealous superiors (North African governor Musa for Tariq, Cuban governor Velázquez for Cortés) who ordered them to abandon their expeditions, and both were chastised afterward (Tariq being briefly imprisoned, Cortés being named only 1st Marquis of the Valley of Oaxaca, a minor honor) despite their fabulous successes.

Cortés may have mimicked many aspects of Tariq's Iberian conquest, but there was one key difference: Cortés left one boat untouched. As Cortés told his men:

> As for me, I have chosen my part. I will remain here, while there is one to bear me company. If there be any so craven, as to shrink from sharing the dangers of our glorious enterprise, let them go home, in God's name. There is still one vessel left. Let them take that and return to Cuba. They can tell there how they deserted their commander and their comrades, and patiently wait till we return loaded with the spoils of the Aztecs.

Leaving one boat was a stroke of genius, as it forced each of Cortés's men to choose whether to remain, while also creating an intense social pressure not to be one of the craven few to return to Cuba. Having chosen to remain, his men were then psychologically committed to the mission in a way that wouldn't have been possible if Cortés had just sunk the whole fleet and held them hostage.

Cortés's decision to scuttle his ships is often described as an example of commitment. However, this decision didn't commit Cortés himself in any meaningful way, since he planned to remain in Mexico for some time and had little need for the fleet. Rather, it committed *his men* not to return to Cuba, by making it impossible for all of them to leave and by creating a new incentive (not to appear craven) to remain.

Committing others to do what you want has an obvious appeal. If you are able to change what strategies others can play and/or change their payoffs, as Cortés did, then you can gain a strategic advantage by inducing them to take actions that benefit you. This idea applies even to games that we play with ourselves, such as when we make commitments so as to be able to resist self-destructive impulses.

## Example: Resisting Temptation

> It is a man's own mind, not his enemy or foe, that lures him
> into evil ways.
>
> —*Buddha*

> Take me and bind me to the crosspiece half way up the
> mast; bind me as I stand upright, with a bond so fast that I
> cannot possibly break away.... If I beg and pray you to set
> me free, then bind me more tightly still.
>
> —*Odysseus to his men as they approached the Sirens, in*
> *Homer's* Odyssey

Like me, you may spend your days mostly chained to a desk. Have you ever noticed that you tend to eat and snack *more* in the office than when you are up and about and, presumably, might really need the extra energy? One obvious reason is the easy availability of snacks. More subtle, but just as real, is the effect of sedentary work on our ability to resist temptation.

The scientific journal *Appetite* recently featured a research report on chocolate consumption by those engaged in a computer task.[3] During breaks in that task, subjects were free to take as many chocolates as they wanted from a bowl that was prominently displayed.

First, though, each subject was asked to engage in either fifteen minutes of brisk walking or fifteen minutes of quiet contemplation. Besides the obvious effect of burning calories, exercise is generally regarded as a virtuous act. It's natural then to expect that subjects who exercised would allow themselves a little more vice and consume extra chocolate. In fact, those who exercised actually consumed less chocolate on average, 15.6 grams, compared to the 28.8 grams consumed by those who engaged in quiet contemplation.

How could this be? The leading theory appears to be that exercising affects the mix of chemicals in your brain, suppressing your appetite and cravings for things like chocolate. When you choose to exercise, then, you are playing a game with your "future self,"[4] by changing your future self's desire to eat chocolate. In my own case, for example, the only time I can go running most days is in the morning, while most of my snacking opportunities come in the afternoon. "Morning David" never wants to run, but is willing to do so if that will stop "Afternoon David" from snacking so much. As long as Afternoon David prefers not to snack after a run, my snacking problem is no big deal. Anticipating that Afternoon David will snack *unless* I exercise, Morning David hits the trails.

Sometimes the solution is not so easy. Consider the problem of losing weight. How can you (i.e., your present self) incentivize yourself (i.e., your future self) to make a real effort to lose weight? Just eating less today won't be enough, since your future self may then just splurge and gain all the weight back again. In a 2006 *Forbes* magazine article,[5] economists Ian Ayres and Barry Nalebuff envisioned a new sort of business offering a novel solution to this problem: "weight-loss bonds." Dieters would pay $1,000 for a weight-loss bond and then, as long as they met prespecified dieting goals, would receive an above-market rate of return. Ayres and Nalebuff's business would make money as some dieters "defaulted," while dieters would benefit by having more of an incentive to keep off the pounds.

A weight-loss bond is a way for your present self to incentivize your future self to stick to a diet. In this sense, buying a weight-loss bond is very much like Cortés's decision to sink his fleet, a way for one player (Cortés, your present self) to commit another player (Cortés's men, your future self) to do what it wants.

### HOW IS COMMITMENT RELATED TO "MOVING FIRST"?

Commitments are only effective when made early enough (and visibly enough) to have an impact on what others choose to do. Thus, anyone who commits needs to "move first," in the sense of committing before others make their decisions. But what people typically mean when they speak of "moving first" is the notion of moving earlier or more quickly. The rest of this chapter (and the Game-Theory Focus section that follows) explores what moving first really means, from a strategic point of view.

## Influence the Timing of Moves

Git thar fust with the most men.

—*General Nathan Bedford Forrest, legendary Confederate cavalry commander, on the key to his battlefield success*

"The early bird gets the worm." That may be true for birds but, in business, there isn't always an advantage to being first into a new market space. As marketing professors Gerard Tellis and Peter Golder note in *Will and Vision: How Latecomers Grow to Dominate Markets* (2001): "Market pioneering is neither necessary nor sufficient for long-term success." Indeed, many companies that are widely believed to be pioneers were in fact late arrivals to their categories: Kodak in cameras (their 1888 entry was preceded by the daguerrotype in 1839), Procter & Gamble in diapers (their Pampers launch in 1961 was preceded by Johnson & Johnson's Chux in 1934), Xerox in photocopiers (it faced a crowded field of about thirty copying machine manufacturers when it entered in 1959), and Apple in personal computers (its 1976 launch came eighteen months after the MITS Altair).

These firms dominated their markets for years not because they did things *first*, but because they did things *best*. Consider Apple. The first affordable personal computer wasn't the Apple I but the Altair from Micro Instrumentation and Telemetry Systems (MITS). In July 1976, the very month that Apple I launched, *BusinessWeek* magazine ran a story titled "Microcomputers Catch on Fast,"[6] in which it referred to MITS as the "IBM of home computers" and reported that MITS's "early lead has made its design the de facto industry standard." Indeed, the Altair had a clear lead both in hardware (its method of transferring data within the computer, the S-100 bus, quickly became the industry standard) and in software (its operating system, Altair BASIC, was Microsoft's first ever product). But few have ever heard of the Altair because MITS pursued a losing business strategy. The Altair was sold unassembled, as a hobbyist kit, and the user interface was extremely limited, essentially just a bunch of switches and lights. This left the door open for Apple to enter with a pre-assembled personal computer featuring an easy-to-use video display. Indeed, being "late to the game" may actually have been to Apple's advantage, since it could learn from the Altair's flaws when designing the Apple I.

Other times, being first is absolutely essential. Nowhere is this more true than in military battle, where the side that positions itself first (on more advantageous ground) often enjoys a decisive advantage. Of course, moving an army quickly is also quite hazardous, as it can strain supply lines and force a commander to drive blind. This may help explain why only the rarest commander seems capable of translating the need for speed in battle into a sustained strategic advantage. When one of these rare commanders comes along, however, he can alter the course of an entire war. Nazi Field Marshal Erwin Rommel, master of the art of tank warfare during World War II, is one example of such a Game-Changer.

Even better may be Confederate General Nathan Bedford Forrest. Civil War historians Shelby Foote and Bruce Catton have said of Forrest: "The Civil War illuminated only two men of military genius. One was Abraham Lincoln. The other was Nathan Bedford Forrest"; and,

FIGURE 1   Game-Changers Erwin Rommel and
Nathan Bedford Forrest

"Forrest used his horsemen as a modern general [does] motorized infantry.... Not for nothing did Forrest say the essence of strategy was 'to git thar fust with the most men.'"

Forrest entered the war as an uneducated private but left as the Confederacy's most feared general. Union General William Tecumseh Sherman once wrote to his commander, General Ulysses S. Grant: "I will order them to go out and follow Forrest to the death, even if it costs ten thousand lives and breaks the Treasury. There will never be peace in Tennessee until Forrest is dead!" Certainly, the man was fierce. During the war, he purportedly had thirty horses shot out from under him but personally killed thirty-one Union soldiers. "I was a horse ahead at the end," he said.

Forrest's ability to position and reposition his cavalry forces quickly (and fearlessly) was key to his ability to defeat superior Union forces time and time again throughout the war. In 1864, for instance, Union General Samuel Sturgis descended on Forrest at Brice's Crossroads in northern Mississippi with a force nearly twice as large. Uncowed, Forrest advanced aggressively while Sturgis was still on the move, engaging and defeating Sturgis's cavalry before the rest of his army could arrive. Unhorsed, the Union forces were then blind to Forrest's moves. When he assaulted a bridge near the back of their position, Union troops feared the worst, that Forrest had found a way to attack them from behind. Sturgis ordered a general retreat which quickly descended into a chaotic, panicky rout. The ensuing chase stretched through six counties, until finally Forrest's men grew too tired to ride down any more fleeing Union troops, more than 1,500 of whom were taken prisoner.

In any battle, whether in business or in war, the combatants' desire to win creates an adversarial dynamic that spills even into the question of who moves first. In a military battle, generals like Nathan Bedford Forrest may race to be the first to claim the high ground, while in business, savvy firms may wait in hopes of prompting someone else to pioneer a new market. Not all games are like that. Sometimes, as in the following example, everyone agrees who ought to move first.

## Example: Vacation Rental By Owner (VRBO)

HomeAway Inc. is the giant of the vacation rental by owner (VRBO) market, with many popular websites including HomeAway.com, VRBO.com, VacationRentals.com, and BedandBreakfast.com. Home Away.com alone offers "more than 325,000 vacation rentals to choose from." With so many properties, however, it is difficult to ensure the accuracy of every listing. Not surprisingly, some owners slip through the cracks with deceptive descriptions of their rental properties. There is even a term for this phenomenon: "SNAD" ("significantly not as described").

HomeAway.com user "vixen25" described her own SNAD experience on HomeAway's community discussion board:

Everything in the ad turned out to be grossly exaggerated. The house was not within walking distance to the beach (it was a 15 minute drive at best). There were no large sauce pans to cook with (despite the house being rented for 14 guests). The dishwasher was broken. And on and on.[7]

At least vixen25 had a place to stay. In a November 2011 piece titled "Vacation rental scams are a growing problem,"[8] consumer advocate Christopher Elliott told the even more unfortunate story of Tania Rieben. Ms. Rieben had wired $4,300 for a six-week Maui condo rental to an owner she found through VRBO.com, only to discover that that person's VRBO.com account had been hacked and her money had been lost to a scam. Worse yet, neither the owner nor VRBO.com took any responsibility for her loss, leaving her with no money . . . and no vacation.

From a game-theory perspective, the fundamental problem here relates to the timing of moves: renters must pay to reserve a property *before* they can verify whether it has been described accurately. This gives unscrupulous property owners an incentive to post deceptive listings. Fortunately, a new site called Airbnb.com has burst onto the scene with a radical new business model that solves this problem. Under Airbnb's approach, renters are not charged until twenty-four hours *after* their stay begins. This gives renters a chance to inspect the property and ensure that deceptive property owners aren't paid.[9] Anticipating this, owners have an incentive to describe their properties truthfully, while scammers have nothing to gain from hacking into property owners' accounts.

Everyone wins in a system like this, in which renters don't need to worry about scams or SNADs. The only potential loser is HomeAway, whose system fails to engender the same level of automatic trust and whose market is therefore potentially vulnerable to invasion by the Airbnb model. Perhaps this explains why upstart Airbnb already had a $1 billion valuation in July 2011.[10] HomeAway's CEO, Brian Sharples, recently told the *Wall Street Journal* that he isn't worried ("Those guys are really good, [but] not as good as we are"),[11] but perhaps he should be. Even better, perhaps he should learn from what works at Airbnb and find a way to let HomeAway property owners "move first," so that vacationers can be even more confident in the quality of HomeAway rentals.

## Sometimes, Moving First Isn't Possible

In October 2000, Sony launched the Playstation 2 (PS2), the best-selling videogame console of all time. (As of January 2011, 150 million PS2 devices had been sold, along with 1.5 billion games.) A year later, Microsoft and Nintendo launched their own highly anticipated "sixth-generation" consoles, known as the Xbox and GameCube. Each of these competitors faced a fundamental decision: whether to pursue a "me-too" strategy and attempt to match or better the PS2 in terms of graphics and immersive experience, or to design a device with different strengths that would compete less directly with Sony. In effect, Microsoft and Nintendo had to decide whether to enter the same computation-intensive niche that Sony had already carved out, or strike out in their own directions.

Whether it's a good move for (say) Microsoft to enter Sony's turf depends on how Sony is likely to respond. Will Sony launch an all-out price war, selling the PS2 at a loss to ensure that Microsoft also loses money? Or will Sony be more accommodating and set a profitable price at which both can peacefully coexist? It's not worth entering Sony's territory if doing so will trigger a price war, but otherwise there is room enough for both Sony and Microsoft to make plenty of money sharing the hardcore gamer market. Figure 2 describes this Xbox Entry Game in the form of a game tree, where Sony naturally decides whether to launch a price war *after* Microsoft decides whether to enter with a similar device.

**FIGURE 2**   Game tree for the Xbox Entry Game

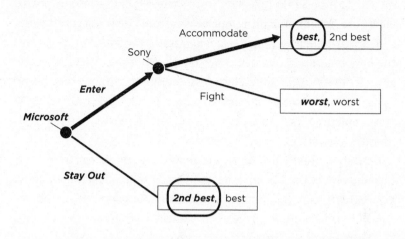

## HOW TO READ A GAME TREE

A game tree provides a convenient way to summarize each player's strategic options in games in which players move sequentially. For instance, figure 2 illustrates how (i) Microsoft moves first, deciding whether or not to enter the hardcore gamer market, and (ii) if Microsoft enters, Sony must then decide whether or not to fight. The game tree also shows how each player ranks the possible outcomes of the game, from best to worst, under the convention that the first-mover's payoff is listed first. For instance, "Microsoft Enter + Sony Accommodate" is the best possible outcome for Microsoft but only the second-best outcome for Sony, while "Microsoft Enter + Sony Fight" is the worst possible outcome for both.

Returning to our discussion of the Xbox Entry Game, Sony could have potentially deterred Microsoft from entering its market by committing, ahead of time, to fight any entry with an all-out price war. Such a commitment would effectively allow Sony to move first, from a strategic point of view, and induce Microsoft to stay out. Unfortunately for Sony, there was no way to *credibly* commit to such a threat. Why not? Once the Xbox launches, there is no hope of inducing Microsoft to take it off store shelves.[12] Thus, the only possible future benefit of fighting a price war would be if Microsoft were sufficiently punished that it decided to stay away from Sony's turf in the *next* generation of videogame consoles. But that's several years away, a "lifetime" in Sony's business. Moreover, without the ability to harvest sufficient profits now to plow into next-generation R & D, Sony could find itself in a risky and vulnerable position in the next product generation.

Thus, Sony really had no way to move first and deter Microsoft's introduction of the Xbox.[13] Sony remained profitable, but not nearly as profitable as it would have been if only Microsoft had chosen to stay out of the console business, or launched a different sort of console that didn't compete so directly with the PS2.

In all of the games that we've considered so far, it matters who moves first. But there are other situations in which the likely outcome of the game doesn't depend on the timing of moves. The most famous such game is the Prisoners' Dilemma.

## The Prisoners' Dilemma

The police have arrested two criminals on charges that carry a prison term of up to five years, but strongly suspect that they also committed a worse crime (say, armed robbery) that carries a term of up to twenty years. The police interrogator puts them in separate cells and says to each, "It's time for you to confess to the armed robbery. How long you stay in prison will depend on who confesses. If you're the only one to confess, I will let you walk free today because of your cooperation. Otherwise, you'll spend five years behind bars if neither of you confesses, ten years if both of you confess, and twenty years if you're the only one not to confess."

Figure 3 illustrates the players' payoffs (in terms of jail time) in each possible outcome of the game, using a so-called payoff matrix.

FIGURE 3   Payoff matrix for the Prisoners' Dilemma

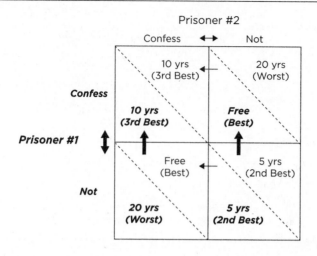

These diagrams will be used throughout the book so, before proceeding, let me describe how to read them. (While potentially confusing at first, payoff matrices will become easy to read and understand once you get used to them.)

## HOW TO READ A PAYOFF MATRIX

A payoff matrix is a quick and easy way to summarize players' incentives in a game, as well as to draw strategic connections between games that may at first glance seem to have little in common. Every payoff matrix shows (i) who the players are, (ii) each player's available moves, (iii) the possible outcomes depending on players' chosen moves, and (iv) how players rank these possible outcomes, i.e., players' payoffs. In addition, I will often use "incentive arrows" to illustrate whether and how players' incentives depend on what others do.

1. *Players*: Payoff matrices are typically used to describe two-player games. One player is called "Row player" while the other

**FIGURE 4**    Generic payoff matrix (without incentive arrows)

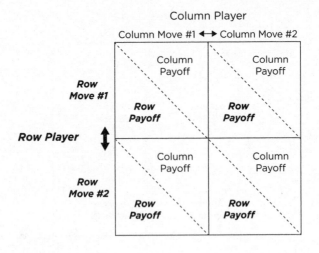

is called "Column player." Their names appear, respectively, on the left and at the top of the diagram. For further clarity, all terms relating to Row player are ***italicized and bolded***, while all terms relating to Column player are in regular print.

2. *Moves*: Each row of the matrix corresponds to a possible move of Row player, while each column corresponds to a possible move of Column player.

3. *Outcomes*: Each box of the matrix corresponds to a possible outcome of the game. In the 2 x 2 example shown in figure 4, there are four possible outcomes.

4. *Payoffs*: Depending on players' chosen moves, each player will get a payoff. By definition, a player's payoff captures everything that he/she cares about vis-à-vis the outcome of the game.[14] This allows us to rank all of the possible outcomes from each player's point of view. Each box of the payoff matrix shows how both players rank the outcome corresponding to that box, with Row player's ranking in the bottom left trian-

gle and Column player's ranking in the upper right triangle
within the box.

5. *Incentive arrows*: A payoff matrix can be helpful for visualizing
   each player's incentives in the game, i.e., how each player's pre-
   ferred move (a.k.a. "best response") depends on the other player's move. To illustrate such incentives, I use up–down arrows
   to show Row player's incentives and left–right arrows to show
   Column player's incentives.

Returning to the Prisoners' Dilemma, whose payoff matrix is shown
in figure 3, note that each prisoner has a unilateral incentive to confess, regardless of the other's move. (If the other prisoner confesses,
confessing reduces your own sentence from twenty to ten years. If the
other prisoner does not confess, confessing allows you to avoid prison
entirely.) That is, each prisoner has a "dominant strategy" to confess.
However, if both confess, both get a longer sentence (ten years, the
third-best outcome) than if neither confessed (five years, the second-
best outcome).[15]

---

*A "dominant strategy" is a move that maximizes a player's
own payoff, regardless of others' moves, holding
others' moves fixed.*

---

Princeton mathematician Albert W. Tucker[*] created the story of the
Prisoners' Dilemma in 1950 as an example for a lecture on game theory
to psychology students. Since then, others have generalized the game
to apply to situations with many players. In its most general form, the

---

[*]Tucker (1905–95) stood out for his generous spirit, passion for mathematics education (he helped found the AP Calculus exam), and outstanding PhD advisees. In
1950, the same year that Tucker coined the term "Prisoners' Dilemma," one of those
students—John Nash—submitted a PhD thesis that would later earn the Nobel Prize
in Economics.

Prisoners' Dilemma is defined as any game having the following two features:

1. *Each player has a dominant strategy*, a move that maximizes that player's own payoff regardless of others' moves. (Each prisoner has a dominant strategy to confess.)
2. *All players are worse off* when they all play their dominant strategies, compared to when each plays some other strategy. (Both prisoners are worse off when both confess, compared to when neither confesses.)

### COMMUNICATING AND/OR MOVING FIRST DOESN'T HELP

A key feature of Tucker's classic tale is that the prisoners are isolated in separate cells, unable to communicate or observe whether the other has confessed. But these features aren't essential to the dilemma. To see why, consider a variation of Tucker's original story in which the police interrogator brings both prisoners to the same cell and makes a slightly different speech: "It's time for you both to confess to the armed robbery. I'm going to leave you alone for ten minutes. Talk it over. When I come back, you will each have just one chance to confess. First, I'll ask you [Prisoner #1] whether you confess, then you'll leave the room. Then, I'll ask you [Prisoner #2] whether you confess."

It's easy to imagine what the prisoners will talk about during their ten minutes. Prisoner #2, especially, needs to convince Prisoner #1 that he won't confess once Prisoner #1 has left the room, if only Prisoner #1 also doesn't confess. But whatever Prisoner #2 may promise, Prisoner #1 knows that "words are wind" and that Prisoner #2 will prefer to confess once the time comes. Anticipating this, Prisoner #1 will also confess when given the chance. They'll still both go to jail for ten years, notwithstanding their chance to communicate and Prisoner #1's chance to move first.

# The Greater Significance of the Prisoners' Dilemma

In competition, individual ambition serves
the common good.

—*Russell Crowe (as John Nash) in* A Beautiful Mind *(2001),*
*completely missing the point of the Prisoners' Dilemma*

The Prisoners' Dilemma is, without a doubt, the most studied and widely cited of all games. Yet some feel that the Prisoners' Dilemma has received more than its fair share of attention. Richard H. McAdams (no relation to the author), the Bernard D. Meltzer Professor of Law at the University of Chicago, made this point in the context of game theory and the law:

> Legal scholars are nearly obsessed with the Prisoners' Dilemma, mentioning the game in a staggering number of law review articles (over 3,000), but virtually ignoring other equally simple games that offer equally sharp insights into legal problems.... In particular, the need for coordination is as pervasive and important to law as the Prisoners' Dilemma, such as in constitutional law, international law, property disputes, traffic, culture, gender roles, and many other topics. Further, the profession's over-focus on the Prisoners' Dilemma unnecessarily contributes to the divide between Law & Economics and Law & Society scholars, all of whom might find some common ground in exploring coordination games.[16]

The "coordination games" that Professor McAdams mentions here are those, naturally enough, in which players have an incentive to coordinate their moves. For instance, in traffic, we all benefit when everyone drives on the same side of the road. I agree that coordina-

tion games are important but, nonetheless, the Prisoners' Dilemma
has earned its pride of place.

First of all, many important and vexing real-world games are
Prisoners' Dilemmas. In business, for instance, competition itself
can be a Prisoners' Dilemma (see chapter 3). Indeed, even the most
basic aspect of business, the transaction, can be viewed as a Prisoners' Dilemma (see chapter 5). Fortunately for business, firms long ago
found ways to escape the Prisoners' Dilemma of competition, reducing or even eliminating their incentive to compete, while trusted
institutions have arisen to facilitate transactions.

Perhaps the most important aspect of the Prisoners' Dilemma
is that it presents an eminently *solvable* strategic problem. Indeed,
game theory provides five distinct "escape routes" from the Prisoners' Dilemma,[17] each of which is broadly relevant in many other games
as well:

1. Regulation (chapter 2)
2. Cartelization (chapter 3)
3. Retaliation (chapter 4)
4. Trust (chapter 5)
5. Relationships (chapter 6)

A deeper understanding of the Prisoners' Dilemma can also enrich
the ongoing philosophical and political debate, sometimes carica-
tured as "capitalism vs. socialism," over the proper scope of individual freedom, personal responsibility, and collective action. The
Prisoners' Dilemma encompasses any situation in which individual
incentives conflict with the greater good, so much so that everyone is
worse off when everyone pursues their own self-interest.[*]

In this way, the Prisoners' Dilemma embodies the fundamental
distinction between license and liberty and highlights the need, in

---

[*]When competition takes the form of a Prisoners' Dilemma, it is *not* true that "individual ambition serves the common good." Russell Crowe was wrong.

some situations, to restrict our ability to make certain choices and/ or to increase our personal responsibility for the consequences of our actions. After all, even the most ardent defender of personal liberty can appreciate the damage and chaos of "freedom" run amok, and the importance of institutions that protect our liberty to make a good life for ourselves while ensuring that we don't deny others that same opportunity.

# GAME-THEORY FOCUS 1:
# THE TIMING OF MOVES

The way that games are likely to play out depends critically on what game theorists refer to as the timing of moves. This terminology is a bit misleading, since the player who moves first from a chronological point of view need not be the "first-mover" from a strategic perspective. The chronological order in which players make their moves matters, but so does their observability and capability to pre-commit to how they will play the game. For the sake of clarity, I will focus here on the simplest sort of game, in which two players each make a single irreversible move. There are three possibilities for the timing of moves in such games:

- Simultaneous moves
- Sequential moves
- Commitment moves

## Simultaneous Moves

A game has simultaneous moves if each player must decide what to do without observing the other player's choice, i.e., if the players make their choices in mutual ignorance. The term "simultaneous moves" comes from the fact that mutual ignorance is automatic if both players make their moves at exactly the same moment. However, chronological simultaneity is not necessary for a game to have simultaneous moves in the strategic sense, as illustrated by the following example.

*A game has "simultaneous moves" if each player does not observe the other player's move prior to choosing its own move, i.e., if players make their moves in mutual ignorance.*

## Example: The Battle of the Bismarck Sea

> Our losses for this single battle were fantastic. Not during the entire savage fighting at Guadalcanal did we suffer a single comparable blow.
>
> —*Masatake Okumiya, Japanese staff officer at Rabaul, 1943*

In January 1943, just a year after the attack on Pearl Harbor, Japan's imperial forces were on the defensive. They had just lost Guadalcanal and, after the brutal battle of Buna-Gona, the Allies had landed forces on New Guinea that might soon threaten the important Japanese base at Lae. Desperate to turn the tide, Imperial General Headquarters dispatched Major General Toru Okabe's 51st Infantry Division to convoy to Lae from nearby Rabaul and drive the Allies off the island. Allied aircraft intercepted Okabe's convoy and decimated the fleet of Japanese aircraft that was protecting it, along with a few vessels carrying critical supplies, but most of the convoy made it through. It soon became clear, however, that Okabe's forces could not successfully drive out the Allies without reinforcements and supplies. And so one of the most critical decisions of the Pacific War was made, to send two more Japanese divisions in a make-or-break gamble to retake New Guinea.

These new troops arrived in Rabaul and, like those before them,

FIGURE 5   The second Rabaul-Lae Convoy, March 1943

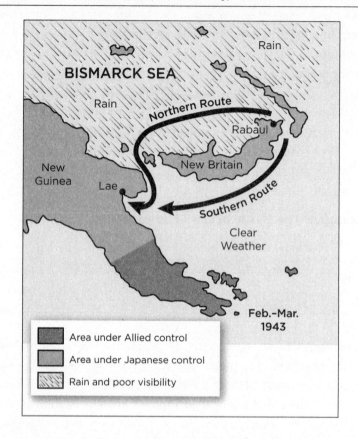

faced the prospect of a treacherous passage through waters within reach of Allied bombers. Worse yet, because of Japanese aircraft losses during the first convoy, the second convoy had little air protection. Indeed, it wasn't a question of whether the second convoy would be bombed on the way to Lae, but of how long they would suffer such attacks. The answer would depend on whether the Japanese convoy took the northern or southern route to Lae, and over which of these routes the Allies sent most of their limited reconnaissance craft.

Retired US Air Force Colonel O. G. Haywood considered this game—a decisive episode in the Battle of the Bismarck Sea—in his

**FIGURE 6**   Possible outcomes of the second Rabaul-Lae Convoy

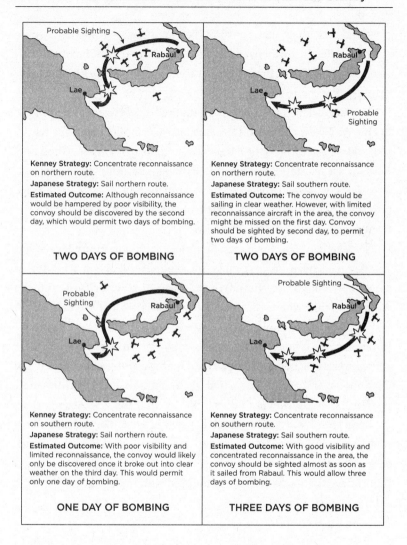

**Kenney Strategy:** Concentrate reconnaissance on northern route.

**Japanese Strategy:** Sail northern route.

**Estimated Outcome:** Although reconnaissance would be hampered by poor visibility, the convoy should be discovered by the second day, which would permit two days of bombing.

**TWO DAYS OF BOMBING**

**Kenney Strategy:** Concentrate reconnaissance on northern route.

**Japanese Strategy:** Sail southern route.

**Estimated Outcome:** The convoy would be sailing in clear weather. However, with limited reconnaissance aircraft in the area, the convoy might be missed on the first day. Convoy should be sighted by second day, to permit two days of bombing.

**TWO DAYS OF BOMBING**

**Kenney Strategy:** Concentrate reconnaissance on southern route.

**Japanese Strategy:** Sail northern route.

**Estimated Outcome:** With poor visibility and limited reconnaissance, the convoy would likely only be discovered once it broke out into clear weather on the third day. This would permit only one day of bombing.

**ONE DAY OF BOMBING**

**Kenney Strategy:** Concentrate reconnaissance on southern route.

**Japanese Strategy:** Sail southern route.

**Estimated Outcome:** With good visibility and concentrated reconnaissance in the area, the convoy should be sighted almost as soon as it sailed from Rabaul. This would allow three days of bombing.

**THREE DAYS OF BOMBING**

1954 classic article, "Military Decision and Game Theory."[1] As he explained it, the Japanese would face from one to three days of bombing, depending on which route they took and which route the Allies patrolled. See figure 6, which can be interpreted as an (unlabeled)

payoff matrix, with the Allies (under General George Kenney) as the Row player.

In the end, the Japanese took the cloud-covered northern route and the Allies also focused their patrols there. The Japanese suffered two decimating days of Allied bombing runs. All eight Japanese transports were lost, along with four destroyer escorts and nearly 3,000 troops. This crushing defeat marked a decisive moment in the Pacific War. Never again did the Japanese attempt to reinforce Lae by sea, and without that option they were unable to stem the Allied tide. Less than a year later, Lae had fallen, Rabaul lay crippled and broken, and the Japanese themselves were crouched in a purely defensive posture. In a real sense, the Battle of the Bismarck Sea marked the beginning of the end of the Pacific War.

The Battle of the Bismarck Sea was a game on many levels, but I've focused here on just one: the Japanese decision whether to send their convoy to the north or south of New Britain, and the Allied decision where to send their reconnaissance aircraft. What was the timing of moves in this game? Obviously, the Allies didn't know exactly when the Japanese convoy would depart, so their reconnaissance began before the Japanese put to sea. Since the Japanese couldn't detect Allied reconnaissance craft from their port at Rabaul, however, it didn't matter that the Allies made their decision first from a chronological point of view. The Japanese still had to choose their own route in ignorance of the Allies' move. Since the Allies and the Japanese made their decisions in mutual ignorance, the game therefore had "simultaneous moves" from a strategic point of view.

# Sequential Moves vs. Commitment Moves

A stealthy meeting Thursday night between former
presidential rivals Barack Obama and Hillary Clinton is
no longer a secret, but what they discussed—and whether
they said the words "vice president"—remains a mystery.

—*ABC News report, June 6, 2008*[2]

In August 2008, John McCain was in a tough spot. Rumors were swirling that conservative elements within the Republican Party had pressured the presidential candidate to drop his good friend, Senator Joe Lieberman, from his short list for vice president. Making matters worse, the possibility of an Obama–Clinton ticket seemed quite real. Barack Obama and Hillary Clinton, fierce adversaries in the Democratic presidential primaries, had snuck off to a secret meeting in June where, Senator Dianne Feinstein reported to ABC, "They both left laughing." Senator Chuck Schumer, a close friend and ally of Clinton, was also quoted in mid-June as saying, "She has said if Senator Obama should want her to be vice president and thinks it would be best for the ticket, she will serve, she will accept that."

McCain's camp had been hoping for Obama and Clinton's relationship to remain frosty, since that might mean many of Clinton's female supporters would sit out the election. Indeed, were Obama not to choose Clinton, McCain might hope to attract some of those female votes himself, especially if he were to pick a female running mate such as Kay Bailey Hutchison (Texas senator), Carly Fiorina (businesswoman), Sarah Palin (Alaska governor), or Condoleezza Rice (former secretary of state). So, in choosing his own running mate, McCain desperately needed to know: would Obama pick Clinton, or not?

Fortunately for McCain, the Democrats held their national convention a week before the Republicans that year, forcing Obama to reveal his choice (Senator Joe Biden) before McCain had to finalize his own. Furthermore, given all the lead time between when McCain secured the Republican nomination and his own convention, there was no inherent reason why McCain couldn't have multiple vice presidential candidates fully vetted and ready to go, some best suited if Obama chose Clinton and others best suited if he did not.[3] Since McCain was capable of observing and responding to Obama's choice, their game of choosing vice presidential nominees did *not* have simultaneous moves. Rather, Obama was the first-mover and McCain was the last-mover.

How games with a first-mover and a last-mover are likely to play out depends critically on one more factor: whether the last-mover is able to commit ahead of time to how it will respond to the first-mover's choice. To emphasize this, I will use different terminology for the timing of moves when the last-mover is capable or incapable of committing to how it will respond. In particular, a game has "sequential moves" if the last-mover cannot commit ahead of time to how it will respond, while it has "commitment moves" if the last-mover can so commit.

---

*A game has "sequential moves" if (i) some player ("last-mover") can observe and respond to the other player ("first-mover")'s move and (ii) the last-mover cannot commit ahead of time to how it will respond to what the first-mover does.*

---

*A game has "commitment moves" if (i) some player ("last-mover") can observe and respond to the other player ("first-mover")'s move and (ii) the last-mover can commit ahead of time to how it will respond to what the first-mover does.*

---

In 2008, McCain had an overriding incentive to pick whichever running mate would maximize his chance of winning the presidency. In other words, McCain's response to Obama's choice was dictated by

McCain's own desire to win. Consequently, while McCain moved last, he did *not* have last-mover commitment power, so the game itself had sequential moves. McCain might have changed the timing of moves by announcing and credibly committing to his intended candidate ahead of the Democrats' national convention. Since Obama could observe and respond to such an announcement, McCain would then have been the "first-mover" from a strategic perspective. However, this is not the same as making a commitment move.

For a game to have commitment moves, the last-mover must be able to commit to *any* response, even one that hurts. Consider the threat of corporal punishment of naughty children, mostly out of favor nowadays but widely used by generations of loving parents. "This is going to hurt me more than it hurts you" was the dreaded refrain just before a spanking or whipping. The threat of corporal punishment by a loving parent is a commitment move, since (i) the parent can observe and respond to the child's behavior (parent is "last-mover") and (ii) the parent can commit to respond to the child's behavior by doing things that s/he doesn't want to do (parent has "last-mover commitment power").

What's essential here is that the parent commits to how s/he will *respond*, which is very different from tying one's hands. Knowing that they will be spared a spanking if only they behave, children have an incentive to stay out of trouble. Tying your hands in this context would mean committing always to beat your children, no matter how they behave. Unlike the *threat* of punishment for misbehavior, such a cruel commitment would have no deterrent effect on the child. Recognizing this, law enforcement authorities arrest parents who beat their children without reason (as first-movers), but grant some latitude to parents who beat their children after misbehavior (as last-movers).

# Changing the Timing of Moves

Since game outcomes depend on the timing of moves, it should come as no surprise that players routinely take steps to change the timing of moves to their advantage. There are three basic ways to do so.

### 1. CHANGE OBSERVABILITY

One way to make your move observable is to cultivate third parties (e.g., rating agencies, auditors, consumers on feedback forums) who can credibly *report* on what you do. Or, to keep your move secret, *"signal jam"*[4] by making statements and taking steps that could be consistent with more than one course of action. For instance, one way for a politician to keep her intentions secret would be to quietly deputize friends and associates to speculate about a wide variety of possible plans. That way, even if someone should leak the truth, the media and her opposition may not even notice.

### 2. CHANGE CHRONOLOGICAL TIMING

One way to move first is to impose an *artificial deadline* on yourself. Another way to move first is to grant the other player *"inspection rights"*—the ability to verify your move before they make their own choice. For instance, offering a money-back guarantee gives customers the right to return a product that they do not enjoy, in effect allowing them to move last. Or, to move last yourself, acquire the *flexibility* to change your move until the bitter end, by taking preparatory steps that keep all your options open. For instance, a supplier who is bidding for a contract could commit to beat any competitor's offer. If so, the buyer will be sure to give the supplier a chance to observe and respond to any competing bid. Similarly, on eBay, "sniping" software allows bidders to submit offers at the very last moment. As such, sniping software allows bidders to observe as much as possible about others' bids before committing to their own.

### 3. CULTIVATE THE LAST-MOVER'S COMMITMENT POWER

One way to make a commitment credible is to tie the outcome of the current game to something else that's bigger and more important, for instance, by invoking *personal honor* and/or committing to a *future relationship*. If you renege on your commitment today, you will tarnish your honor and/or lose all future benefits from the relationship. As long as your incentive to cheat today is small enough, such steps can allow you to credibly commit not to cheat. Absent honor or relational concerns, you could also sign an *enforceable contract* that specifies damages should you renege on your commitment.

# CHAPTER 2

# Invite Regulation

In 1968, ecologist Garrett Hardin coined the term "tragedy of the commons" to refer to any game in which individuals can freely choose how much to exploit an exhaustible resource. Each player has a dominant strategy to use as much of the resource as possible but, when everyone does so, the resource is overexploited and everyone suffers. (So, the tragedy of the commons is an example of the Prisoners' Dilemma.) The idea of the tragedy of the commons, that resource exploitation is a game, has since been widely applied to understand many so-called common resource problems,[1] such as pollution, overfishing and habitat destruction, highway traffic, and spam email.

Hardin famously argued that the only way to solve the tragedy of the commons is to restrict individuals' rights to use common resources. This perspective has had a profound impact, justifying major government interventions—championed by those on both sides of the political spectrum—that restrict free usage rights. On one hand, liberals invoke Hardin's argument to justify direct government control of common resources. On the other hand, conservatives invoke the same argument to justify privatization of the commons.

Moves toward direct government control are often referred to as "greater regulation," while moves toward privatization are called "deregulation." From a game-theory perspective, however, government control and privatization are similar, as they consolidate the decentralized and often informal control that communities exercise over their common resources into the hands of a single third party, chosen by the government.

Much of the political debate between liberals and conservatives is then really about which third party is likely to be a better (or less bad) steward of the commons and its fair use: a government agency or a for-profit business. In fact, there is a third option: rather than stripping communities of their right to control their own common resources, empower them to work together to use the commons in their own collective interest. Elinor Ostrom won the 2009 Nobel Prize in Economics for demonstrating how such *community self-regulation* can avoid the tragedy of the commons.

## What Is "Regulation"?

The idea of regulation is to change players' payoffs, to give them the incentive to make different decisions than they otherwise would make. The Code of Federal Regulations weighs in with over 150,000 pages of rules, but that's just the tip of the iceberg of government regulation. Every law passed by Congress, every action taken by the executive branch, and every ruling by the judiciary "regulates" the governed.

Moreover, government isn't the only regulator. Administrators in schools, parents in families, and trendsetters in society all "regulate" others by influencing their options and incentives. Perhaps above all, firms are at heart regulatory bodies, in the game-theory sense of incentivizing others to change their behavior, as they seek to motivate employees to work harder, entice customers to buy more, and so on.

> *Firms are at heart regulatory bodies, in the game-theory sense*
> *of incentivizing others to change their behavior, as they seek*
> *to motivate employees to work harder, entice customers to buy*
> *more, and so on.*

These examples all hinge on an asymmetry of power, as the regulator dictates terms to the regulated, but this isn't essential. Indeed, communities of all kinds routinely regulate themselves—e.g., the member colleges of the NCAA which regulate college sport.

### Example: College Football Violence

American football was invented in the 1870s, as college students at Harvard and a few other schools experimented with new variations of rugby and soccer.[2] Then as now, violent contact was a central element of the game. Indeed, one of the most common ways to bring down a ball carrier in those early days was simply to strike him in the face with a closed fist. As time went on, teams adopted more and more violent tactics, resulting in terrible injuries and even dozens of deaths. It got so bad that President Theodore Roosevelt[3] personally intervened in 1905, threatening to ban the sport and triggering a series of reforms that would lead to the creation of the National Collegiate Athletic Association (NCAA) in 1906.[4]

Regulation was essential to tame college football because escalating violence was the natural strategic evolution of the sport. Consider the infamous "flying wedge," a V-shaped offensive formation used nowadays mainly by riot police, that can literally run over the defending side. Harvard was the first to employ the flying wedge, in a surprise move against archrival Yale in 1892. The next year, however, neither team had an advantage—and both suffered worse injuries—as both employed the wedge.

## THE FOOTBALL VIOLENCE GAME

We can think of Harvard and Yale's decisions whether or not to employ the flying wedge as a game itself. Each team's best outcome is to be the only one to use the wedge and worst outcome is to be the only one not to do so. Further, players on both teams are better off if neither uses the wedge than if both do so, since serious injury is less likely without this dangerous tactic in play. (These payoffs are illustrated in figure 7.) Note that each team has a dominant strategy to employ the wedge, but both teams are worse off when both do so. Consequently, this game is a Prisoners' Dilemma.

Why is employing the wedge a dominant strategy for each team? Consider Harvard, the Row player. Recall that, by definition, a dominant strategy is any move that maximizes a player's payoff (i.e., that is a "best response"), regardless of others' moves. Recall also that the up–down arrows in the payoff matrix capture Harvard's incentives, pointing to the row(s) corresponding to Harvard's best response to each of Yale's possible moves. To see that Harvard has a dominant strategy, then, it suffices to note that all of the up–down arrows point to the same row ("Wedge").

Why is Harvard's best response always to employ the wedge? Let's think through the possibilities. First, what if Yale employs the wedge? Harvard also then has an incentive to do so, to avoid its worst outcome, in which only Yale employs the wedge, and instead get its third-best outcome, in which both teams use the wedge. Next, what if Yale does not employ the wedge? Harvard still has an incentive to use it, to get its best outcome in which only Harvard employs the wedge instead of its second-best outcome in which neither team uses the wedge. Thus, employing the wedge is Harvard's dominant strategy.

Once the NCAA established stricter rules, to outlaw the most violent tactics, these incentives changed. With rules in place, any team that used illegal tactics would be penalized, putting them at a disadvantage. Of course, teams still had the *option* to use the wedge, but as long as referees did their job, teams had no *incentive* to do so. In other words, stricter rules changed the payoffs in the Football Violence

FIGURE 7    Payoff matrix for the Football Violence Game

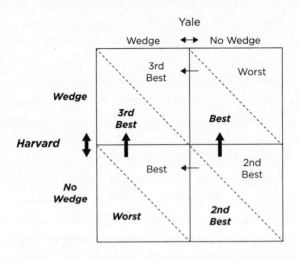

Game, transforming it into a less violent game in which each team had a dominant strategy not to employ the flying wedge.

> *Regulation allows players to escape the Prisoners' Dilemma by changing their payoffs, and thereby changing their incentives.*

Regulation is essential to redress some social ills. For instance, fisheries regulation plays a crucial role in preserving fish stocks, while the "priority review vouchers" described in the END Amendment example below could spur new treatments for neglected tropical diseases. However, unintended consequences can sometimes do more harm than the intended good. For instance, who would have thought that banning cigarette advertisements in 1970 would have the secondhand effect of *increasing* the number of people who smoke?

## Example: The Cigarette Advertising Ban

> Millions are spent, year after year, to promote cigarettes.
> No one will argue that these expenditures are to no
> significant purpose. They are significant.
>
> *—Rosel Hyde, chairman of the Federal Communications*
> *Commission, in testimony before Congress, 1969*[5]

World War I was a defining moment for tobacco in the United States. As General John J. Pershing wrote to Army Command in 1917, "You ask me what we need to win this war. I answer tobacco as much as bullets. Tobacco is as indispensable as the daily ration; we must have thousands of tons without delay." Soldiers' daily tobacco ration may have helped to win the war but, more important for the tobacco industry, it also created a robust demand for cigarettes after the troops came home.

Cigarette manufacturers competed aggressively throughout the mid-twentieth century to capture this growing market, with iconic characters such as the Marlboro Man, memorable taglines such as "Lucky Strike: It's Toasted," and reassurances about the health-fulness of cigarettes such as "20,679 Doctors Say 'Luckies Are Less Irritating'" and "Not One Single Case of Throat Irritation Due to Smoking Camels . . . Camels: *Costlier* Tobacco."[6] As the truth about the health hazards of cigarette smoking became more widely known, public health authorities became concerned that such advertisements were luring people into becoming smokers. The Federal Communications Commission (FCC) responded in 1967 by mandating that any television station airing cigarette ads must also air public-service announcements highlighting the dangers of cigarette smoking.

Congress responded further with the landmark Public Health Cigarette Smoking Act of 1970, requiring all cigarette packages to be labeled with a warning ("The Surgeon General Has Determined That Cigarette Smoking Is Dangerous to Your Health") and banning cigarette advertisements on American radio and television. In exchange, the FCC stopped airing its anti-smoking public messages and cigarette manu-facturers were granted immunity from future federal lawsuits.

What many Americans don't know is that most of this legislation—with the exception of the warning-label requirement—was actually proposed by the tobacco industry. It's fairly obvious why tobacco executives would want immunity from federal lawsuits, which could bankrupt or dismantle their companies and even land them in prison. But why ban their own TV and radio ads?

First and most obviously, such a ban could help stop growing momen-tum toward even more injurious regulation.[7] Second, halting their own ads also put an end to the FCC's anti-smoking campaign. The *New York Times* reported in 1970: "The tobacco industry [was] convinced that the anti-smoking messages were hurting business more than the ads were helping it and that to drop both would bring a net gain."[8] So, collectively banning all their ads was a win for the tobacco industry.

Prior to the advertising ban, each firm had an individual incen-tive to continue to air its own ads. To see why, consider the stylized Cigarette Advertising Game, whose payoff matrix is shown in figure 8. In this game, each firm—Philip Morris (Marlboro brand) and R. J. Reynolds (Camel brand)—has a dominant strategy to advertise its own brand. The reason is that the benefit of advertising your own brand, in terms of stealing market share, exceeds the damage done to the overall market by the FCC anti-smoking message that your ad will trigger to run as well. As long as both firms run TV ads, how-ever, their efforts to steal market share largely cancel out, leaving both worse off due to the FCC's anti-smoking campaign. Since both firms are worse off when both play their dominant strategy of adver-tising, this Cigarette Advertising Game is yet another example of the Prisoners' Dilemma.

The notion that firms might actually benefit from not advertising may strike some readers as odd, so let's dig a bit deeper. If cigarette advertising in 1970 really was a Prisoners' Dilemma, we would expect to see both a drop in advertising and an *increase* in profits after the advertisement ban, as firms move from the worse outcome in which

**FIGURE 8** Payoff matrix for the Cigarette Advertising Game

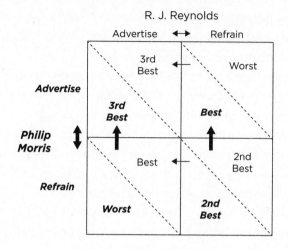

all advertise to the better outcome in which none advertise. Indeed, this is exactly what happened. How do we know?

The cigarette ban created a unique opportunity for what social scientists refer to as an event study. If we look at advertising expenditures and industry profits just before and just after the ban went into effect, any difference can be interpreted with some confidence as being a consequence of the ban itself. (By contrast, longer-term trends may be due to a combination of many factors.) Economics professor James L. Hamilton seized on this opportunity in a 1972 article titled "The Demand for Cigarettes: Advertising, the Health Scare, and the Cigarette Advertising Ban."[9] His main finding was that "advertising expenditures were 20–30 percent smaller" in 1971 than in 1970, while "for the first six months of 1971, industry earnings were 30 percent larger than for the same period in 1970."

The tobacco industry earned this windfall—freedom from federal lawsuits and a 30 percent boost in industry-wide earnings—by understanding the real game of cigarette advertising better than its regulators. Anti-smoking advocates at the FCC assumed that cigarette ads must be convincing people to smoke. Otherwise, why would Big Tobacco be spending millions of dollars on them each year? But Big Tobacco was actually a collection of intensely competitive firms, who advertised mainly to steal smokers from one another and only secondarily to induce people to start smoking.

If only regulators had understood the real game, they could have continued to use the Prisoners' Dilemma of cigarette advertising to their advantage. As it stood before the advertising ban, the tobacco industry was fully—and "voluntarily"—subsidizing an anti-smoking campaign that was killing its business.[10] If only regulators had understood the effectiveness of their own policy, they could have doubled down and mandated matching messages in magazines and other major media as well. Doing so would have been an essentially costless way to speed up the decline of cigarette smoking in the United States.

## Example: The Eliminating Neglected Diseases (END) Amendment

> Some of the highest-leverage work that government can
> do is to set policy and disburse funds in ways that create
> market incentives for business activity that improves the
> lives of the poor.
>
> —Bill Gates, speaking at the 2008 World Economic Summit

After malaria, schistosomiasis (also known as snail fever) imposes the greatest health burden on the developing world of all so-called "neglected tropical diseases." Snail fever is endemic in dozens of developing countries, infecting more than 207 million people, 85 percent of whom live in Africa, and placing more than 700 million people at risk. The disease is caused by a small parasitic worm that enters through the skin upon contact with infected water, and subsequently disperses its larvae back into the water supply via bloody urine. (The name "snail fever" comes from the essential intermediate role that certain snails play in the worm's life cycle.) While rarely fatal, snail fever is a chronic illness that can damage internal organs and, in children, impair growth and cognitive development. Fortunately, snail fever is easily treated. Just one tablet per year of the drug praziquantel is enough to clear worms and prevent transmission. Even better, praziquantel only costs about 20–30 cents per treatment.[11] Despite this low cost, the World Health Organization is chronically short of funds and dependent on donations from pharmaceutical companies for its supply of praziquantel. The largest donor in recent years has been Merck, who committed in 2007 to supply 200 million doses over ten years. These donated drugs have undoubtedly eased the suffering of millions of infected people but, unfortunately, they are not nearly enough to push back against the disease in a meaningful way, to free people from exposure to the worms.

This changed dramatically in January 2012, when industry and the global health community came together to issue the London Declaration, an ambitious road map to eradicate or restrict a host of hitherto neglected tropical diseases.[12] A highlight of the London Declaration was Merck's commitment to increase tenfold its donation of praziquantel, to 250 million doses per year.[13] Hailed as a "game-changer," Merck's donation will provide enough drugs to treat every single person infected with snail fever.

Eradication of snail fever is now in sight, if only health officials can find a way to deliver and administer these drugs effectively. Merck is also helping with this delivery effort, by developing a new formulation of praziquantel more suitable for children.[14] (Existing pills are not kid-friendly, being large and bitter to the taste.) This is well and good, even touching. But what's in it for Merck's shareholders? Wouldn't they be better served if Merck's scientists were assigned to alternative projects with a greater monetary payoff? Actually, it turns out, there is serious money to be made by those who find new ways to treat neglected diseases.

In 2007, Congress passed the Eliminating Neglected Diseases (END) Amendment to incentivize pharmaceutical companies to develop new drugs to combat neglected diseases. Any company that develops such a drug receives a "priority review voucher" that gives its owner the right to skip to the front of the line for Food and Drug Administration review of any blockbuster drug in its portfolio.[15] As Bill Gates noted at the 2008 World Economic Summit in Davos:

> Under a law signed by President Bush last year, any drug company that develops a new treatment for a neglected disease like malaria or TB can get priority review from the Food and Drug Administration for another product they've made. If you develop a new drug for malaria, your profitable cholesterol-lowering drug could go on the market a year earlier. This priority review could be worth hundreds of millions of dollars.

The END Amendment could have a transformative effect on human welfare, if it encourages for-profit pharmaceutical firms to pursue (or at least allow their scientists to pursue, as side projects) game-changing new treatments and vaccines for neglected diseases. Unfortunately, what we have seen so far is just the lowest-hanging and lowest-impact fruit. Indeed, as of late 2012, only one priority review voucher had yet been granted, to Novartis for the antimalarial drug Coartem. But Coartem was first developed in 1996 and has been used internationally for over a decade, meaning that Novartis received a priority review voucher while doing nothing to actually advance the treatment of neglected diseases. (Coartem qualified Novartis for a priority review voucher, perversely, because it had not previously been approved by the FDA for use in the United States.)

Fortunately, better things seem likely to come of the END Amendment. Several new drugs are already in the pipeline and may qualify for priority review vouchers in the near future, including a formulation of moxidectin to treat onchocerciasis (a.k.a. river blindness).[16] Even more exciting, for-profit firms that are developing fundamental new approaches to fight disease are looking *first* at neglected diseases, in addition to the usual suspects that afflict the developed world.

Consider NanoViricides, a small but ambitious nanotechnology firm developing an antiviral technology that uses nanomachines that pose as human cells but then envelop and "eat" any virus that binds to them. NanoViricides launched with a focus on big-market opportunities such as HIV and influenza. Recently, however, they announced the results of animal studies showing progress toward a new treatment of dengue fever.[17] This is big news, as dengue fever infects 50–100 million people annually but has no effective treatment other than "supportive treatments" like rehydration, and can worsen into a potentially fatal condition known as dengue hemorrhagic fever.

NanoViricides is a "penny stock," a firm whose future success is so uncertain that its stock trades (under the symbol NNVC on OTC BB) for mere pennies per share, a firm for whom every dollar counts right now. Indeed, as founder Anil Diwan noted in 2011, his firm had only spent $14 million over the past several years, in contrast with competitors who are spending $25 million per quarter just to develop one or two drugs. It's safe to say, then, that NanoViricides would never have focused *so early* on dengue fever had there not been the prospect of a valuable priority review voucher should they ultimately be successful.

By providing that inducement to think and work on neglected diseases, the END Amendment has focused for-profit industry on what had long been a not-for-profit problem. The end result is more innovation, and more collaboration, to save and improve more lives in neglected parts of the world.

# GAME-THEORY FOCUS 2:
# STRATEGIC EVOLUTION

From the sport's invention in the 1870s to the creation of the NCAA in 1906, college football gradually evolved into an ever more violent game. Why did it take so long? Why weren't extreme tactics like the flying wedge introduced earlier, when they could have provided just as much advantage to whoever used them? The answer is obvious: the value of these tactics had not yet been discovered. Once they were introduced, however, their success prompted quick imitation and dispersion throughout the sport. Genetic mutations in the biological world are much the same—new tactics being tried that, if successful, spread throughout a broader breeding group. In this way, the evolution of "strategic ecosystems," such as the world of college football, is fundamentally similar to Darwinian evolution,[1] with new strategies in place of mutations.

Some people may feel uncomfortable drawing this parallel between human and animal behavior. Indeed, in most economics textbooks, human decision makers are treated as being quite sophisticated and "rational," insofar as they (i) form a coherent belief about the world and (ii) make a choice that maximizes their own welfare given that belief. But what if you take the dimmer view that, most of the time, most of us have no clue and just stumble through with whatever seems to work well enough? You're not alone.

There is a long tradition within economics of recognizing and incorporating the reality of non-optimizing behavior into models of human decision making. Indeed, Herbert Simon won the 1978 Nobel Prize in Economics for work (dating from the 1940s and 1950s) on "bounded rationality." Simon's economic insights were grounded in the idea that people routinely possess limited information about the decisions they face and the observation that, given such limitations, we tend to develop heuristics or "rules of thumb" to achieve a *good enough* result.

## Experimentation and Search

How does one determine what's good enough? One way is to experiment. Experimentation (also called "economic search") is one of the most important branches of economic theory. Indeed, the 2010 Nobel Prize in Economics went to Peter Diamond, Dale Mortensen, and Chris Pissarides for their work on search frictions in markets. However, one of the best explanations of economic search that I've heard was by a non-economist, Edward Conard, who earned a fortune working at Bain Capital alongside Mitt Romney. During an interview with the *New York Times*, Conard expounded on the best, most rational way to find a spouse:

> Set aside a bit of time for "calibration"—dating as many people as you can so that you have a sense of what the marriage marketplace is like. Then you enter the selection phase, this time with the goal of picking a permanent mate. The first woman you date who is a better match than the best woman you met during the calibration phase is, therefore, the person you should marry.[2]

This sort of search is great for some things, like choosing your favorite burrito joint. But don't get carried away and ignore the *strategic* aspects of search. For instance, suppose that you adopted

Conard's mate-search protocol but, like most of us, lacked his millions of dollars in the bank. Suspecting that you felt no true love when you made your vows, your spouse will naturally invest less in the relationship and your marriage will be weaker than it could have been. There's a reason why we're endowed with the capacity for unreasonable love: so that we can stick together and work as a team, even in hard times. That's why it's far more rational—and strategic—to marry for love.

This talk of finding the right spouse is interesting and fun, but also a bit limiting. Let me pull us back, then, to the more basic problem of finding the right strategy in games. If you had the chance to play a game over and over, you might experiment a bit and settle on whatever tends to work best. But what if you only had one chance to play? You could still look around at what others are doing and use that information to help guide your decision.

## Strategic Evolution

"Strategic evolution" refers to the tendency of players in a strategic ecosystem to adapt/evolve their strategies, knowingly or unknowingly, toward whatever is a best response to others' current strategies. For instance, the college admissions ecosystem is populated by high school seniors applying to college and admissions officers deciding who to accept, not to mention parents and alumni who care about admissions and third parties like the College Board (which administers the SAT and AP exams) and *US News & World Report* (which ranks colleges based largely on admissions outcomes). Interestingly, this ecosystem is *not* in equilibrium, as players constantly adapt to others' strategies in a complicated, interconnected dance.

## Example: College Admissions and the College Board

> The more any quantitative social indicator is used for
> social decision-making, the more subject it will be to
> corruption pressures and the more apt it will be to distort
> and corrupt the social processes it is intended to monitor.
>
> —Donald T. Campbell, president of the American
> Psychological Association, 1976

In 1926, about 8,000 students sat for the first ever administration of the Scholastic Aptitude Test (SAT). By 2011, nearly three million students took the test.[3] According to the College Board, "The SAT tests the reading, writing and mathematics skills that students learn in school, and that are critical for success in college and beyond." The SAT may or may not test skills that are truly "critical for success," but one thing's for certain: the SAT is central to the formula that *US News & World Report* uses when ranking the "best" American colleges.

As *US News* has emerged as a powerhouse over the years, steering more and more of the best high school applicants to apply to the "best" colleges on its list, college admissions officers have responded by putting more emphasis on SAT scores in their admissions decisions. (By admitting students with higher SAT scores, a college can rise in the *US News* ranking and attract better applicants next year.) Responding to this, high school students and their parents put more effort into maximizing their SAT scores. Indeed, the wealthiest students now routinely hire superstar tutors who charge as much as $300 an hour,[4] or even pay others to take the test for them.[5]

Responding to growing concerns about the fairness of the SAT, a number of schools have joined an emerging "SAT-optional" movement that allows students to decide whether to submit their scores when they apply.[6] Of course, only those with high scores will choose to reveal them, so that the *reported* average SAT score at an SAT-optional school could be much higher than the *true* average SAT score among students at that school. According to one admissions dean interviewed anonymously by the *Washington Post*, this allows SAT-optional schools to rise in the *US News* ranking by, effectively, failing to disclose low scores.[7] If so, *US News*'s ranking of SAT-optional schools may skew them too highly, potentially undermining the validity and accuracy of the ranking itself.[8] How *US News* might respond to such a development is unclear, but one natural possibility would be to de-emphasize the SAT itself, putting more emphasis on other measures of academic success such as Advanced Placement (AP) courses.

The Advanced Placement program has long been one of the great success stories of American education. Founded in the 1950s by a group of inspired educators (including Albert Tucker, who coined the term "Prisoners' Dilemma"), the Advanced Placement Committees in subjects ranging from calculus and physics to English literature and Russian language have created exams that even critics tend to applaud as consistent and valid indicators of academic success. It's not surprising, then, that politicians and school administrators have latched onto the AP curriculum in their quest to improve America's public schools.

In 2000, the US secretary of education and the president of the College Board announced their goal to offer ten AP classes in every high school in the United States. Since then, school districts across the nation have invested heavily to add the AP curriculum. William Lichten, professor emeritus of physics, engineering, and applied science at Yale University, has argued that this push has been ineffective at many schools, especially lower-performing ones:

> Rather than being prepared for advanced college courses, many times graduates from struggling high schools need remedial courses on arrival at college. A simple question apparently left unasked [by those promoting expansion of AP to low-performing schools] is how high schools that have graduates in need of remedial college coursework can be expected to teach students to succeed on AP exams designed to determine if students may bypass

introductory college courses for advanced college courses. The error in reasoning is palpable.[9]

The College Board is unfazed by such criticism. Its minority-oriented "Get with the Program" promotion plainly advocates the AP curriculum, even for those who do not plan to attend college: "AP isn't just for top students or those headed for college. AP offers something for everyone," and, "If college isn't in your plans, AP is still a great choice.... No matter what you decide to do in life, the lessons you learn in AP can boost your self-confidence and put you on a path toward success."[10]

Unfortunately, given the AP failure rates at many schools, it's unclear what sort of confidence boost the program offers. For example, the Jacksonville, Florida, *Times-Union* reported that "In the 1995–96 school year, 29 students attended an advanced placement American history class at Andrew Jackson High School and took a final exam to qualify for college credits. No one passed."[11] More broadly, the available evidence suggests that, outside of schools with selective admission, the push to add AP classes to public schools has borne surprisingly little fruit. In the Philadelphia school system in 2006, for example, 179 AP classes were offered in forty-one public schools. Selective "exam schools" (such as Masterman High School, touted as "one of the best high schools in the tri-state area") performed well, with passing rates in some subjects above the national average. But of the thirty-two schools reporting low average SAT-V scores (between 313 and 408), twenty-seven reported passing rates of less than 10 percent, and the highest passing rate was only 33 percent.[12]

Even without the political push to expand AP into low-performing schools, students at the best schools have already long been under intense pressure to take Advanced Placement exams to get a leg up in college admissions. Twenty years ago, only the most accomplished high school students took even one Advanced Placement exam. Since then, participation in the Advanced Placement program has exploded, and it's now the norm for students at elite colleges to have taken several AP exams. Top colleges have responded by designating some AP exams as unworthy of college credit,[13] and requiring higher scores to qualify for credit on the exams that do count,[14] diluting the value of students' effort to take and pass AP exams in the first place.

The explosion of the AP program is certainly good news for the College Board, which now earns more revenue from APs than from all other sources combined (including the SAT and the PSAT). Such growth has generated its own detractors. Some educators contend that "AP courses are not remotely equivalent to college-level courses" and that "the AP classroom is where intellectual curiosity goes to die."[15] These critics worry that, in today's hypercompetitive college landscape, top students may focus on taking as many AP courses as they possibly can rather than on the intellectual excitement of the journey into more challenging material that AP can provide.[16] If so, the AP experience won't be nearly as satisfying or meaningful as it once was, and it's just a matter of time before a whole new backlash begins, unleashing new waves of changes and adaptations into the strategic ecosystem of high school and college admissions.

On and on that system may evolve, perhaps never reaching an equilibrium with a stable status quo. That may seem depressing, but it's actually good news if you are sufficiently game-aware to understand the forces driving such change. Whether you are a parent, a school administrator, or an influencer like *US News* or the College Board, you can then anticipate and position yourself advantageously vis-à-vis where the system is headed. Isaac Asimov summed up this perspective in 1978, in an essay entitled "My Own View":

> It is change, continuing change, inevitable change, that is the dominant factor in society today. No sensible decision can be made any longer without taking into account not only the world as it is, but the world as it will be.... This, in turn, means that our statesmen, our businessmen, our everyman must take on a science fictional way of thinking.[17]

Asimov's "science fictional way of thinking" is the forward-looking view of an open and imaginative mind, to understand the forces shaping society and anticipate where we are headed. We know this way of thinking and being, more simply, as game-awareness.

# Evolutionary Stability

Evolutionary biologists were among the first to grab on to the game-theory notion of strategic evolution, as a way to understand the dynamics of Darwinian evolution. Indeed, taking a page from game theory, evolutionary biologists routinely refer to "evolutionarily stable strategies" (ESS) as a concept to help understand the possible stable outcomes of an evolutionary process.[*] As the college admissions example above and the side-blotched lizard example below show, however, some evolving systems may never reach a stable state.

### Example: More Than Lizards

Strategic evolution can be counterintuitive. Charles Darwin's crucial insight was that, through the processes of natural selection and sexual selection, animals are shaped by the games they play. Indeed, using formulas derived from game theory, scientists can predict just how animal populations are likely to evolve over time. One such study appeared in *Nature* in 1996.[18] Behavioral ecologists Barry Sinervo and C. M. Lively observed an isolated population of side-blotched lizards, whose males are genetically programmed with one of three (fixed and heritable) mating strategies. These three types of males are noticeably different in several ways, including the colors of their throats:

---

*A strategic ecosystem is "evolutionarily stable" if (i) the system has entered equilibrium, i.e., has stopped changing, and (ii) this equilibrium is robust to small perturbations, i.e., any small change in the system will tend to be automatically reversed/corrected by further evolutionary change.

- *Blue-throats*: Blue-throated males are monogamous. They keep their partners close.
- *Orange-throats*: Orange-throats are large and aggressive. Their mating strategy is to support a harem of many females.
- *Yellow-throats*: Yellow-throats are the dandies of the lizard world. Effeminate in appearance (females also have yellow throats), these lotharios sneak into orange-throat territory to woo unattended females.

When a female decides with whom to mate, she is essentially choosing the color of her sons' throats and thereby determining how successful her sons will be, relative to other females' sons, at mating in the next generation.

Interestingly, this game among females is fundamentally the same as the children's game Rock Paper Scissors, when we view orange-throats as Rock, yellow-throats as Paper, and blue-throats as Scissors. Orange-throat sons are most successful against a population consisting mainly of blue-throats (Rock beats Scissors), while blue-throats are most successful against yellow-throats (Scissors beats Paper), and yellow-throats do best against orange-throats (Paper beats Rock).[19] Consequently, no single mating strategy tends to become dominant, since once one strategy (say orange-throat) gains prominence, another (yellow-throat) will become more successful and grow as a fraction of the overall population.

After studying these lizards' ecological environment to determine the (reproductive) payoffs in this game, Sinervo and Lively derived formulas to predict the trajectory of the population mix between orange-, blue-, and yellow-throats. Interestingly, they predicted that the population mix would *not* stabilize but, instead, "orbit" around the ESS every several years. And, indeed, after going back to observe the lizard population year after year, that's exactly what they found.

Those lizards provide a potent reminder of how games shape those who play them. It's natural to look at the games in our lives and assume that we chose to make them that way. But, more often than not, the opposite is true. Rather than changing the games that we play, we let those games change us. At least, that's true of most people, those who haven't yet become game-aware. They are like puppets, dancing on a string held by the laws of game theory. Freedom to chart one's own strategic destiny comes only to those game-aware enough to rise above the game, and determined enough to change the game to their own strategic advantage.

# CHAPTER 3

# Merge or "Collude"

Every contract, combination in the form of trust
or otherwise, or conspiracy, in restraint of trade or
commerce among the several States, or with foreign
nations, is declared to be illegal.

—*Sherman Antitrust Act, 1890*

Barbed wire reshaped the American West after its invention in the 1870s, as farmers finally had an effective way to enclose their land. The only problem for this new industry was the simplicity of the product and the ease with which firms could enter the market. Indeed, from 1873 to 1899, as many as 150 companies manufactured barbed wire. As the market matured, however, just a handful of the most successful companies remained. These firms, whose founders were known as the Big Four, were aggressive competitors, each with an incentive to offer more attractive pricing to secure a larger share of the market.

Figure 9 illustrates firms' payoffs in this Competitive Pricing Game, using two of the leading firms (Barb Fence Co. and Southern Wire Co.) as representative players. Each firm's best outcome is to be the only one offering a low price, since it can then capture the lion's share of the market, and worst outcome is to be the only one offering a high price. Of course, if both charge the same price, both prefer a high price over a low price.

Note that each firm has a dominant strategy to price low. Why? Consider Barb Fence. If Southern Wire prices high, Barb Fence pre-

**FIGURE 9   Payoff matrix for the Competitive Pricing Game**

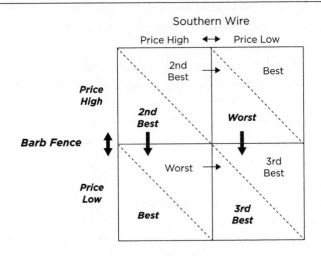

fers to price low to capture the lion's share of the market (best outcome) rather than to split the market at a high price (second-best outcome). If Southern Wire instead prices low, Barb Fence again prefers to price low, though now to avoid its worst outcome in which Southern Wire gets the lion's share of the market. Since both firms prefer to price low, regardless of the other's move, both have a dominant strategy. However, when both play that dominant strategy and price low, both are worse off than if both had set prices high. So, the Competitive Pricing Game is a Prisoners' Dilemma.

The Big Four were trapped in this state of intense competition, but not for long. In 1899, under the leadership of John Warne Gates, they merged to form the American Steel and Wire Company, instantly transforming barbed wire from one of the most competitive industries to one of the most profitable. It's easy to imagine how Gates got the idea to push for a merger. In the late nineteenth century, waves of consolidation had spread throughout American industry, allowing industrialists such as Andrew Carnegie (steel) and John D. Rockefeller (oil) to acquire unprecedented control over massive swaths of

American business. Yet such concentrated power prompted a back-lash, as presidents William McKinley and Theodore Roosevelt made "trust-busting" a central mission of their administrations.

*Cartelization allows players to escape the Prisoners' Dilemma of competition by merging into a single entity to look after their collective interest.*

Nowadays, all large mergers are reviewed by federal regulators (at the Justice Department, Federal Trade Commission, and/or Federal Communications Commission, depending on the industry) and rou-tinely blocked if these regulators fear they could hurt consumers by decreasing competition.[1] Consequently, colluding to escape the Pris-oners' Dilemma of competition isn't as easy today as it was in the gilded age of Carnegie and Rockefeller.

## Collusion Isn't Always Bad

The term "collusion" brings to mind smoky backroom deals that stifle competition and harm consumers. Yet collusion can be viewed more broadly as cooperation among one set of players that just happens to harm another set of players. Indeed, collusion may even be desirable from a social welfare point of view, if we especially value the welfare of the colluding group, or if there are still other benefits from their collusive activities that outweigh whatever harm they cause.

*"Collusion" is cooperation among one set of players that harms another set of players. Collusion is not always a bad thing.*

In the diamond market, for instance, the De Beers cartel for a long time blocked the entry of competing diamond distributors and kept diamond prices high. This undoubtedly hurt De Beers' potential competition, but what about consumers? As I argue below in "Diamonds Aren't Forever Anymore," De Beers' high (and stable) prices might have actually been good for consumers, as they enhanced the diamond's symbolic value.

Other business cartels bring even more clear-cut benefits to consumers. For instance, consider an industry in which firms primarily compete on research (say, to develop new pharmaceutical drugs) but in which the greatest research success comes when *pairs* of firms work together. Once the first pair of firms teams up in this way, they will produce new drugs more quickly and thereby harm the other firms in the market. If left unchecked, these "colluders" could wind up driving everyone else out of business and, ultimately, harm consumers through higher prices. Anticipating this, however, other firms are likely also to team up. In the end, the market could be transformed from one in which individual firms compete to one in which pairs of firms compete. Such an industry-wide transformation would increase the pace of discovery, making consumers better off as well. Recognizing such potential benefits, regulators at the US Department of Justice (DOJ) and the Federal Trade Commission (FTC) use less stringent guidelines when evaluating "collaboration" among competitors than when evaluating mergers of competitors.[2]

More broadly speaking, such efficiency arguments have prompted antitrust authorities to adopt a more permissive attitude toward several sorts of business activity that could conceivably be construed as "collusion." For instance, recognizing how intellectual property (IP) rights can be used both to promote and to stifle innovation in areas like drug and device development, the DOJ and the FTC evaluate exclusive IP licensing deals differently from other sorts of "exclusive dealing."[3] Some groups are even exempted by law from antitrust scrutiny, including labor unions, farm cooperatives, and most major-league sports.[4]

The sports exemption and farm cooperative exemption have come under some fire in recent years,[5] but overall, the general tendency in

American antitrust has been for the list of exemptions to grow over time, as Congress, the Supreme Court, and regulators recognize the potential efficiency benefits of certain sorts of business combinations. Indeed, as I argue below in "Dialing for Dollars," recent developments that have rocked the world of charitable giving may call for yet another antitrust exemption, this time for charities.

## Example: Diamonds Aren't Forever Anymore

In 1477, Archduke Maximilian of Austria commissioned the first known diamond engagement ring for Mary of Burgundy, his betrothed. A new trend was born among Europe's royalty and wealthiest elites. Diamonds were exceedingly rare and expensive at that time, being found only in a few river deposits in India. But then, in 1869, a massive diamond lode was discovered in Kimberley, South Africa, prompting a rush of 50,000 miners to the area. Diamond prices began to fall, a price decline that would continue for decades.

Lower prices meant that diamonds were finally attainable for the masses,[6] who promptly imitated the royal custom of gifting diamond rings upon engagement. Yet as more and more "regular" people began wearing diamonds, they fell out of favor with the elite and, as they fell out of favor with the elite, they also began to lose some of their luster with the masses. This experience taught the new South African diamond magnates a crucial lesson: falling prices could damage the diamond business. For most products with growing demand, lower prices are essential to spur widespread adoption. For diamond engagement rings, however, it's the diamond's symbolic value that people are paying for. And what's the symbolism of a rock that's going to be worth less and less with each passing year?

To stabilize prices, South African miners had to sell fewer diamonds into the market. Yet no individual miner had an incentive to do so. Indeed, the miners found themselves in a Prisoners' Dilemma, each with a dominant strategy to sell every diamond that he found but all worse off when no one withheld any diamonds from the market. The miners solved this problem of oversupply much as the industrialists of America's gilded age avoided competition, by merging into one giant syndicate under the leadership of legendary De Beers founder Cecil Rhodes.

Cecil Rhodes arrived late to the Kimberley diamond rush, in 1874, with no claim of his own. All he had was a steam-powered water pump, the rights to which he had acquired from its inventor. As it happened, prospectors at the Kimberley Mine were having a water problem—seepage from the water table grew ever more severe as they dug deeper into the ground—and Rhodes's pump became essential for any miner who wanted to remain in business. As you can imagine, Rhodes charged an arm and a leg for the right to use his pump, quickly amassing sufficient wealth to begin buying up the motley assortment of Kimberley mines himself. Before long, Rhodes controlled the entire South African supply of diamonds and had created a worldwide distribution network (known as the Diamond Syndicate) to channel that supply all the way from mine to market.[7]

As Nicholas ("Nicky") F. Oppenheimer, chairman of De Beers, explained in a keynote address at the Harvard Business School Global Alumni Conference in March 1999:

> Diamonds are the ultimate luxury and yet they are desired and owned by a vast number of people. They are seen as the ultimate gift that lasts forever and has a store of value. But dealing in a complete luxury that lasts forever and has a store of value lays on some very firm disciplines. We at De Beers never dare forget that the material quality of a person's life would not be changed if they never bought a diamond. The purchase of a diamond at engagement is a mixture of commitment, beauty, and store of value—a heady cocktail of emotion and practicality. Certainly, anyone who makes that investment becomes a supporter of the [De Beers-monopolized] single channel marketing, with its aim of preserving value.[8]

Long at the core of De Beers' corporate identity is the notion that De Beers is exempt from the usual laws of business competition. Again, from Nicky Oppenheimer's 1999 Harvard Business School speech:

> I am chairman of De Beers, a company that likes to think of itself as the world's best-known and longest-running monopoly. We set out, as a matter of policy, to break the commandments of Mr. Sherman [i.e., to violate the Sherman Antitrust Act]. We make no pretense that we are not seeking to manage the diamond market, to control supply, to manage prices, and to act collusively with our partners in the business.... Despite all this, we believe that what we do is not only good for us, and all producers of diamonds, but is also in the interests of the consumer.

To exempt itself from the regulatory control of antitrust authorities such as the US Justice Department, De Beers long avoided having any overt American business presence.[9] American antitrust authorities viewed this as skirting the law but, in De Beers' corporate eyes, such regulatory evasion was necessary to maintain the diamond's value proposition and, ultimately, to serve their customers. This sounds suspicious, but there's actually sense to the notion that consumers may be better off without perfect competition in the diamond market.

Consider Oppenheimer's claim that "anyone who makes that investment [i.e., buys a diamond engagement ring] becomes a supporter" of the De Beers cartel. This is clearly true, as anyone who owns a diamond engagement ring prefers for it to remain as valuable as possible—but that's nothing special about diamonds. The same could be said about any durable good, say real estate, where those who currently own a home naturally support any policy that will keep home prices high.

Much more interesting is Oppenheimer's deeper, implicit claim that all consumers are better off with the De Beers cartel in place, even those who have not yet purchased a diamond engagement ring. How could this be? Because what newlyweds are buying is largely a promise, that diamond prices will never fall and hence that their ring will "forever" maintain its symbolic value. A competitive market cannot keep such a promise. But De Beers did, for more than a century.

De Beers' promise-keeping days have come to an end, however, as the diamond market has splintered and fractured in recent years. In 1999, high-end jeweler Tiffany & Co. announced that it was buying a stake in a Canadian mine and would no longer source its diamonds through De Beers. In 2003, Canadian mining group Aber Diamond Corporation purchased the luxury jewelry retailer Harry Winston, giving it storefronts in the United States, Japan, and Switzerland. And so it has continued, with diamond miners partnering with diamond retailers to avoid needing to deal exclusively with De Beers.

This dissolution of De Beers' monopoly position means that there is no longer any single player with sufficient incentive to take the costly actions needed to keep prices smooth and steady. As Guy Leymarie, the managing director of De Beers, remarked to *Fortune* magazine in 2001, "We don't have to go rushing about the world trying to buy every diamond. What is the point of us buying diamonds close to or over our selling prices? It's silly. I'm perfectly happy to market 60%."[10]

Actually, the practice of "rushing about the world" and paying a premium to secure every new diamond was, from the beginning, at the very core of De Beers' business strategy. After all, you can't hope to keep your monopoly power, or your monopoly profits, unless you're willing to pay a premium to absorb all of the competition, wherever they may be. Mr. Leymarie's confession that De Beers no longer even tries to buy up the world's diamond supply means, in effect, that De Beers has given up on maintaining its monopoly.* Indeed, from 2000 to 2005, De Beers' share of the world diamond

---

*Perhaps not coincidentally, the Oppenheimer family recently decided to cash out of the diamond business. See Jana Marais and Thomas Biesheuvel, "Anglo American Ends Oppenheimers' De Beers Dynasty with $5.1 Billion Deal," *Bloomberg*, November 4, 2011.

supply fell from 65 to 43 percent, underscoring the recent sea change in the diamond market.

With De Beers a mere shell of its former self, diamond prices could soon become much more volatile, like other commodities where competition reigns supreme. Unfortunately for diamond sellers, each shock to diamond prices (up or down) will tend to undermine consumers' long-held belief in the "eternity" of a diamond's value and hence its suitability as a symbol of eternal love. Such effects won't be felt overnight—it may take decades—but the ultimate downfall of the diamond engagement ring, when it comes, could occur quite quickly.[11]

Once people get the idea that diamonds are headed for a fall and word of that fear spreads, demand for diamond engagement rings will naturally dip. Prices for diamonds will then fall but, unlike for typical goods, falling prices will not necessarily induce more people to buy. If anything, lower prices will only add credence to people's fears about diamonds as a long-term store of value, further depressing demand and prices for engagement-quality diamonds, in a vicious downward spiral. In the end, lovers everywhere might decide that diamonds aren't really the best or most suitable symbol of their commitment to one another. All because De Beers lost its collusive grip on the diamond market.

### Example: Dialing for Dollars

> It's like a betrayal.... I know I won't donate again. It's like
> they stabbed you in the back. It's terribly wrong.
>
> —*Carol Patterson, American Diabetes Association donor*

Since 2005, an army of millions of Americans has mailed or hand-delivered letters to friends and neighbors, asking them to donate to charities such as the American Diabetes Association and the American Cancer Society. These and other efforts raised a lot of money—just one firm, InfoCision Management, raised $424.5 million for more than thirty nonprofits from 2007 to 2010—but surprisingly little of that money benefited charity. For example, an investigative report by *Bloomberg Markets* magazine in September 2012 revealed that only 22 percent of the proceeds from the Diabetes Association's nationwide neighbor-to-neighbor campaign went to the charity.[12] Everything else went to InfoCision, whose main "contribution" was to call and convince people to donate their time, money (volunteers use their own stamps), and goodwill with friends and neighbors to raise money on its behalf.

Overall, InfoCision kept more than half ($220.6 million) of the $424.5 million that it raised from 2007 to 2010. In some cases, Info-Cision may even have charged charities more than was raised. For instance, according to federal and state filings by the Cancer Society, InfoCision raised $5.3 million for them in 2010 but charged over $5.4 million. In principle, there's nothing wrong with that. As Greg Donaldson, senior vice president at the Cancer Society, explained, "It's certainly not inconsistent for organizations like ours to invest in some loss-leader strategies, to engage people in long-term meaningful relationships."[13] In other words, charities may not get much of the money that InfoCision collects on their behalf, but they do get relationships that can then be milked for years to come.

The problem is that InfoCision deceives donors. For instance, the telemarketer script for the Diabetes Association's campaign instructed solicitors to say that "about 75 percent of every dollar received goes directly to serving people with diabetes and their families, through programs and research," even though the Diabetes Association only got about 22 percent. Worse yet, this deceptive script was approved by the charity. When confronted by *Bloomberg* with this discrepancy, Richard Erb, the vice president of membership and direct marketing at the Diabetes Association, offered "no apologies," saying the association runs many fund-raising campaigns and, overall, about 75 percent of donations fund its programs. "Obviously, if people feel betrayed or that we're not being honest with them, it doesn't make me feel well," Erb told *Bloomberg*, "but the thing is, we're a business. There has never been a time or a place where we said, 'Most of this money is coming to us.'" Except that's exactly what the Diabetes Association instructed InfoCision to tell donors.

### TELEMARKETING-FOR-CHARITY AS A
### PRISONERS' DILEMMA

Why do so many reputable charities—not just the Diabetes Association and Cancer Society but also the American Heart Association, American Lung Association, American Society for the Prevention of Cruelty to Animals, March of Dimes Foundation, and National Multiple Sclerosis Society, among many others—agree to such a small cut of the donations that telemarketers like InfoCision raise on their behalf? The most likely answer is that InfoCision's fees reflect the substantial costs of conducting a telemarketing campaign[14] and that charities themselves would find it even more costly to conduct such campaigns on their own.[15]

The problem with the telemarketing-for-charity business is therefore not that for-profit telemarketers are fleecing charities out of every last dime, but rather the opposite. Apparently, their profit margins are so slim that they feel compelled to adopt aggressive solicitation tactics—including deception—just to squeak by. Moreover, charities themselves have little incentive to pressure telemarketers to tone down these aggressive tactics (at least the legal ones), since this is what allows telemarketers to convert as many people as possible into future donors. Of course, the downside of such tactics is that they may annoy or offend donors and poison the well for other charities that may want to reach those same people in the future.

Charitable giving is, at least to some extent, a zero-sum game. If a telemarketer convinces me to give $100 today to prevent cruelty to animals, that's $100 less that I have tomorrow to give to medical research or any other worthy cause. Recognizing this, every charity has an incentive to go after my money as aggressively as possible. The end result? Willing donors like me are bombarded with countless annoying calls from telemarketers until, eventually, we stop answering the phone and everyone loses. Note that, while each charity has a dominant strategy to conduct aggressive telemarketing campaigns, they all lose as the universe of potential donors is exhausted and even turned off to charitable giving. In this sense, charities are locked in a Prisoners' Dilemma.

## FUND-RAISING CARTELS: A MORE UNITED WAY

Shortly after *Bloomberg*'s exposé of InfoCision's business practices, Senators Richard Blumenthal (D–CT), Herb Kohl (D–WI), and Chuck Grassley (R–IA) called for the Federal Trade Commission, Internal Revenue Service, US Department of Justice, and Consumer Financial Protection Bureau to investigate. A likely result is that Congress will impose new disclosure requirements that force for-profit telemarketers to tell donors just how little of their money will actually go to charity.[16]

How should we expect charities and telemarketers to react to such new disclosure rules? The most immediate impact will likely be on telemarketers' widespread practice of conducting cold calls to identify new donors. By their nature, cold-call campaigns have a low success rate and hence a high cost per dollar raised. Being forced to reveal that high cost will discourage even more people from donating, resulting in an even higher cost per dollar raised, until perhaps no one donates to such campaigns and telemarketers simply give up.

The most likely "winners," as donors grow more resistant to fund-raising middlemen, will be charities like the United Way that are capable of conducting their own nationwide fund-raising campaigns. The United Way traces its history back to 1887 when "a Denver woman, a priest, two ministers, and a rabbi" brainstormed a new way to raise funds to support charitable groups across the Denver area. The real birth of the organization, however, was in 1918 when executives from twelve fund-raising federations took this idea of coordinating (i.e., "cartelizing") the business of charity to the national level, forming the American Association for Community Organizations that would later become the United Way of America.[17]

The United Way's scale has allowed it to raise money, mobilize volunteers, and form partnerships (such as its forty-year relationship with the National Football League) to spread the word about its good works. Other charities might like to duplicate that success, but don't expect to see many United Way copycats anytime soon. Wherever they sit in the world, United Way organizations share a common mission: to empower their local communities in the core areas of education, income, and health. As such, charitable groups that join the United Way can be integrated in ways that enhance their effectiveness.

Charities that focus on narrower issues, such as promoting research to prevent breast cancer or working to end homelessness, would not benefit from merging operations in that way. Such charities need the focus that comes from independent management. Even so, they might still benefit by merging some aspects of their fund-raising operations. That way, at least, they could potentially conduct their own telemarketing campaigns without having to rely on a profit-seeking middleman, at the scale necessary to do so efficiently.

**A NEW ANTITRUST EXEMPTION?**

Such fund-raising coordination could potentially run afoul of US antitrust law.[18] After all, a cartel of charities (let's call it "America Gives") seeking donations over the phone would have a huge advantage over middlemen like InfoCision, who could find it impossible to compete. For charity cartels to be able to enter the telemarketing space and raise their own money, an act of Congress might therefore be needed to establish an antitrust exemption permitting charities to coordinate their fund-raising activities.

Such an exemption has some precedent in the Capper–Volstead Act of 1922, which, among other things, allowed groups of farmers ("farmer-cooperatives") to collude when setting prices. Why did Congress grant farmers this right to collude? In 1848, gold was discovered in California, prompting a mass migration of people to the West. Farming in California and other western states exploded, as did the demand for east–west trade in agricultural products. Then, in 1869, the Central Pacific and Southern Pacific Railroad companies completed the first transcontinental railroad, after which they merged operations to monopolize this vital new trade route.

For the next fifty years, before the Capper–Volstead Act, farmers on both sides of the Rocky Mountains had no real choice but to accept whatever price the railroad monopoly charged.[19] Once allowed to collude, however, farmers could collectively negotiate more favorable prices, lowering the cost of transportation and spurring greater trade and investment in the agricultural sector. In much the same way, most charities today have no real choice but to turn to for-profit telemarketers who, due to economies of scale, can reach potential donors much more cheaply. Once allowed to coordinate their activities, however, charities could work together to increase the effectiveness of their own fund-raising calls, spurring greater giving and charitable activity in the nonprofit sector.

Of course, there are risks in granting any group the right to collude, and Congress will need to be careful when crafting the scope of any "charity exemption" to the antitrust laws. For one thing, safeguards need to be put in place to ensure that charity fund-raising cartels aren't "captured" by some charities to the detriment of others. To see the potential problem, suppose that the biggest charities[20] were to dictate the management of America Gives, the now-lawful fund-raising cartel. These large charities might, consciously or unconsciously, create an unlevel playing field that disadvantages smaller charities.

For instance, suppose that America Gives were to operate on an "equal membership" model, where each member pays a multimillion-dollar membership fee and, in exchange, may employ America Gives' phone banks to make a certain number of calls on its behalf. This sort of arrangement is "fair" to all cartel members, who pay an equal amount to place an equal number of fund-raising calls. But smaller charities would be shut out since they cannot afford the hefty membership fee. Even worse, with the "big guys" out of the market, for-profit options like InfoCision might also disappear, leaving the "little guys" with literally no options at all, since they are too small to launch campaigns on their own.

Such risks can be mitigated by limiting the scope of the antitrust exemption, to disallow cartel practices that (explicitly or implicitly) discriminate against some charities.[21] Of course, such a restriction would create risks of its own, as even well-meaning fund-raising cartels might unintentionally discriminate and run afoul of the law. To decrease such risks, regulators routinely publish so-called "safe harbor" practices that, if adopted, ensure that one is safe from antitrust scrutiny. Defining such safe harbors properly will require careful study and consideration, and I will not presume to propose details here.

What's essential is that charities be allowed to coordinate their fund-raising activities, not just to enable charities to raise money more effectively but to restore donors' faith in charities themselves. Donors have been deceived. Fund-raising cartels could potentially restore their trust.

# GAME-THEORY FOCUS 3:
# EQUILIBRIUM CONCEPTS

Game theorists use the notion of equilibrium to explore how fully rational players with correct beliefs would tend to play any given game. Although irrationality and incorrect beliefs are common in practice, the concept of equilibrium can still be useful as a starting point for thinking about how real-world games are likely to play out. The most famous equilibrium concept is "Nash equilibrium."

## Nash Equilibrium

John Nash, Jr., is one of the most distinguished mathematicians of the twentieth century, the only person ever to win both the Nobel Prize in Economics (in 1994 for "Nash equilibrium") and the Steele Prize in Mathematics (in 1999 for "Nash embedding"). He's also the only mathematician to inspire an Oscar-winning movie, 2001's Best Picture *A Beautiful Mind*, starring Russell Crowe as Nash.

While most famous for his influence on game theory, Nash's contributions to pure mathematics have been, if anything, even more impressive. Indeed, as the Steele Prize committee noted in its award,

"Nash embedding [is] one of the great achievements in mathematical analysis in this century." While the concept of Nash embedding is difficult to state in everyday language—it has to do with making strangely shaped surfaces "fit" into Euclidean space—Nash equilibrium is simple enough for a child to understand.

**Child's definition**: A *Nash equilibrium* is a situation in which everyone is doing the best they can, given what others are doing.

**Formal definition**: A *Nash equilibrium* is a strategy profile such that each player's strategy is a best response to other players' strategies. (A player's "best response" is the strategy that maximizes his/her payoff, given others' strategies.)

### Example: The Lockhorns' Night Out

Leroy and Loretta Lockhorn are a famous cartoon couple—*The Lockhorns* has run since 1968 and been published in 500 newspapers in twenty-three countries—but their marriage isn't an especially happy one. Consider the cartoon shown in figure 10, in which Leroy and Loretta are at a fancy party. Leroy's main objective is clearly to poke fun at his wife's poor driving skills,[*] but his comment also suggests that he and Loretta are playing a game. It's unclear what exactly their strategies and payoffs are in this game, but we can infer two things about it.

First, since "Loretta's driving because I'm drinking," she must prefer the outcome in which Leroy drinks and she drives over the outcome in which Leroy drinks and she does not drive. (It's unclear what exactly would happen if Leroy drank and Loretta did not drive. Would Leroy drive drunk? Would the pair take a taxi home? Loretta's

---

[*]As regular *Lockhorns* readers know, Loretta often gets speeding tickets (in one strip, Leroy asks, "How did you manage to get a ticket from the FAA?") and frequently crashes the car ("Down at the body shop, Loretta is known as the 'supercollider'").

**FIGURE 10**  "The Lockhorns' Night Out"

**"LORETTA'S DRIVING BECAUSE I'M DRINKING, AND I'M DRINKING BECAUSE SHE'S DRIVING."**

choice to drive reveals that she prefers what actually happened over this alternative, whatever it may be.) This is reflected in Loretta's left-pointing incentive arrow in the "Drink" row of the payoff matrix shown in figure 11. Second, since "[Leroy's] drinking because she's driving," Leroy must prefer the outcome in which he drinks and Loretta drives over the outcome in which she drives and he does not drink. This is reflected in Leroy's up-pointing arrow in the "Drive" column of figure 11.

Since Leroy and Loretta are each playing a best response to the other's strategy—drinking is Leroy's best response when Loretta is driving, while driving is Loretta's best response when Leroy is drinking—"Leroy Drinks + Loretta Drives" is a Nash equilibrium.

This might not be the only Nash equilibrium. Indeed, one could read Leroy's comment as suggesting that Loretta is *only* driving because Leroy is drinking (i.e., if he weren't drinking, she wouldn't be driving) and that Leroy is *only* drinking because she is driving. If

FIGURE 11   Incomplete payoff matrix for the Lockhorns' Game

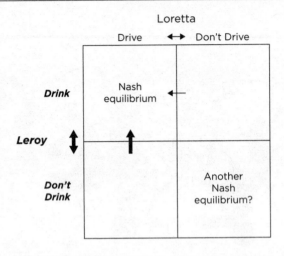

so, "Leroy Doesn't Drink + Loretta Doesn't Drive" would be another Nash equilibrium.[1]

## WHY NASH EQUILIBRIUM CAN BE THE WRONG CONCEPT

The idea of Nash equilibrium seems intuitive, but there are many games in which trusting Nash can get you in trouble. In particular, Nash equilibrium is the wrong concept for thinking about games in which someone moves first. Fortunately, game theorists have devised other concepts, which I will refer to as "Rollback equilibrium" and "Commitment equilibrium," to help us think about such games.

*"Nash equilibrium" only applies to games with simultaneous moves. If some player moves first, we need a different equilibrium concept.*

# Rollback Equilibrium
# and Commitment Equilibrium

Life must be understood backward,
but it must be lived forward.

—*Søren Kierkegaard*

Imagine that the market for flash memory disks is currently dominated by a "Monopolist," who operates a single factory. Another firm ("Entrant") is considering entering the market with one factory, while the Monopolist might expand its own capacity with a second factory. Because of the experience and expertise that comes with incumbency, the Monopolist is capable of building a new factory much faster than the Entrant. However, flash memory disks produced by either company will sell at the same price and can be produced at the same cost, so each firm will enjoy the same per-unit profit margin.

Total profits are highest if there is just one factory, lower but still positive if there are two factories, and negative if there are three factories. (As more factories are built, more quantity is produced, causing prices to fall. The assumption here is that this price-lowering effect is big enough that the market becomes less profitable with each additional factory.) Figure 12 summarizes each firm's profitability, depending on who builds a new factory in this Capacity Entry Game. The resulting payoff matrix is shown in figure 13.

The Monopolist has a dominant strategy *not* to build a second factory. To see why, suppose first that the Entrant were to build a factory. In that case, if the Monopolist were also to build, the resulting excess capacity would make the market unprofitable for them both. So, the Monopolist clearly prefers not to build. What if the Entrant doesn't build? In that case, the Monopolist again prefers not to build, though now the reason is to restrict supply and keep prices high, to squeeze the most out of its continuing monopoly.[2] However, as long as the

FIGURE 12   Profitability in the Capacity Entry Game

| Who Builds? | How Many Factories? | Market Outcome |
| --- | --- | --- |
| No one | 1 | Very Profitable<br>Monopolist Gets All |
| Monopolist only | 2 | Somewhat Profitable<br>Monopolist Gets All |
| Entrant only | 2 | Somewhat Profitable<br>Split with Entrant |
| Both | 3 | Unprofitable for Both |

Entrant anticipates that the Monopolist will not build, the Entrant prefers to enter.

The unique Nash equilibrium of this game is "Monopolist Does Not Build + Entrant Builds," but that's not at all what we would expect to happen. Recall that, by assumption, the Monopolist can build a new factory *faster* than the Entrant. This speed allows the Monopolist, if it so chooses, to break ground and commit to building its own factory before the Entrant can do the same. Once the Entrant sees that the Monopolist is committed to building a second factory, the Entrant prefers to stay out. This leaves the Monopolist in its second-best outcome, "Monopolist Builds + Entrant Stays Out," which is better for the Monopolist than the Nash equilibrium in which it must share the market.

What's going on here? The Monopolist has the capability and incentive to move first. Consequently, the game has sequential moves and the relevant equilibrium concept is not Nash equilibrium but another Nobel Prize–winning idea called Rollback equilibrium.[3] By design, Rollback equilibrium analysis tracks how later-moving players are likely to respond to earlier moves. This allows Rollback equilibrium analysis to capture how first-movers are likely to play when *anticipating* how others are likely to react.

**FIGURE 13   Payoff matrix for the Capacity Entry Game**

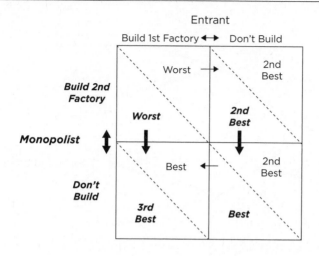

To find the Rollback equilibrium of a sequential-move game, first represent the game and its sequence of moves using a game tree. Figure 14 shows the Capacity Entry Game in game-tree form, with the Monopolist as first-mover. The Monopolist's decision whether to build a second factory comes first and is located at the "root" of the tree. Then, depending on what the Monopolist does, the Entrant chooses which "branch" to go down. Lastly, each "leaf" of the tree gives each player's payoffs in each possible outcome, much as in a payoff matrix. (The Monopolist's payoff is listed first at each leaf.)

We follow Kierkegaard's advice and understand the game backward, starting at the leaves and working down the branches to the root. First, consider the Entrant's decision on the upper branch after the Monopolist has built a second factory. The market will be unprofitable if the Entrant also builds a new factory, so the Entrant prefers to stay out. What about if the Monopolist does not build? In that case, the Entrant prefers to enter, to share a somewhat profitable market. Next, consider the Monopolist. Anticipating how the Entrant will respond depending on whether or not the Monopolist builds a second

FIGURE 14   Game tree for the Capacity Entry Game

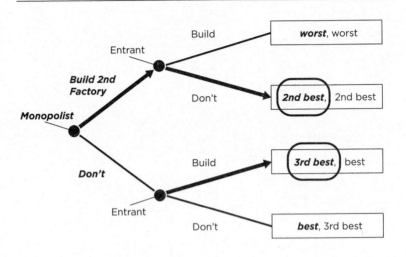

factory, the Monopolist's choice is effectively between the two outcomes pointed to by the thick arrows in figure 14. (The Monopolist's payoffs in these two outcomes are circled.) Obviously, the Monopolist prefers the outcome "Monopolist Builds + Entrant Does Not Build" over "Monopolist Does Not Build + Entrant Builds." So, the Monopolist will choose to build a new factory, as a means to deter the Entrant.

## WHEN ROLLBACK EQUILIBRIUM IS ALSO
## THE WRONG CONCEPT

The central assumption underlying the idea of Rollback equilibrium is that the last-mover is sure to play its best response to whatever the first-mover does. For instance, in the Capacity Entry Game, our Rollback analysis hinged on the assumption that, if the Monopolist commits to build a second factory, the Entrant will respond by staying out of the market. But what if the Entrant had previously convinced the Monopolist that it was going to enter no matter what? Such a commitment, if believed, could cow the Monopolist into not building a new factory after all. In situations like this, when some player moves first but the last-mover has commitment power, we need yet another concept that I will refer to as Commitment equilibrium.[4]

---

*"Rollback equilibrium" applies to games with sequential moves (see Game-Theory Focus 1). In such an equilibrium, the* first-mover *makes whatever move is best for itself, assuming that the last-mover will play a best response to that move.*

---

*"Commitment equilibrium" applies to games with commitment moves (see Game-Theory Focus 1). In such an equilibrium, the* last-mover *makes whatever commitment move is best for itself, assuming that the first-mover will play a best response to that commitment.*

---

Continuing our discussion of the Capacity Entry Game, recall that the Monopolist's only way *as first-mover* to deter entry is to commit quickly to build an additional factory, before the Entrant can do the same. However, even that isn't such a great outcome for the Monopolist, who would be better off not building that second factory. Fortunately, the Monopolist can achieve its best possible outcome *as last-mover* if it can commit ahead of time to how it will respond to the Entrant's move. In particular, suppose that the Monopolist were to commit to the following threat: "If you enter the market with a new factory, I will respond by building my own factory, even though doing so hurts us both." If this threat is believed, the Entrant will view its decision as being between staying out or entering a market with over-capacity. Given these alternatives, the Entrant will choose to stay out, and the Monopolist doesn't even need to build an additional factory to deter entry.

Of course, the trick with commitment moves like this is being believed. Once the Entrant has built a factory, the Monopolist will prefer to accommodate and share the market rather than creating excess capacity that could trigger a price war. Fortunately, there are many ways in which the Monopolist can make such a threat credible, most of which entail changing the game in some way. For example, suppose that the Monopolist CEO's performance is judged primarily on market share rather than profits. If so, the CEO will have a strong incentive to respond to entry by building a second factory, to ensure that his firm remains top dog in the industry. Yet CEO incentives (and CEOs) can themselves be changed by the board of directors. Wouldn't the Monopolist's board prefer to give its CEO an incentive to accommodate, so as to avoid a costly price war?

Perhaps. But all of this discussion has ignored an essential aspect of entry games: *reversibility*. If the physical plant that the Entrant builds to enter the market can be repurposed to compete in another industry, it could make sense for the Monopolist to attack the Entrant as vigorously as possible, incurring some short-term pain in hopes of driving the Entrant away to another market. Doing so would restore its monopoly and, moreover, establish a tough reputation that could deter future entrants from ever challenging its monopoly again.[5]

Even if the Entrant's physical plant cannot be repurposed to another market, it can still be bought up by the Monopolist. Furthermore, bearing in mind that the Entrant will be willing to accept a lower price in bad market conditions, the Monopolist might even find it profitable to build another factory and/or launch a price war itself, for as long as it takes to force the Entrant to sell at a fire-sale price.[*]

---

[*]Of course, antitrust authorities might take notice and punish such a brazenly anticompetitive move.

# CHAPTER 4

# Enable Retaliation

J esse James and Billy the Kid, the two most infamous outlaws of all time, have each just moseyed into Dodge City, Kansas, during the heyday of the American Wild West. The town is holding its breath over what will happen next. Everyone knows that there's not enough room in one town for two such larger-than-life characters, and it's just a matter of time before there's a confrontation. Finally it happens, in the Long Branch Saloon, when Billy and Jesse suddenly come face-to-face, only a few feet apart.[1] In the blink of an eye, both gunslingers whip out their six-shooters and aim true. Both are so good—and so close—that any shot is sure to kill. But neither pulls the trigger. They just stand there, like statues.

Welcome, ladies and gentlemen, to the Mexican Standoff.[2] Each outlaw in this imaginary game is driven primarily by a desire to burnish his own legend as a gunslinger, and only secondarily to survive the day. In particular, each outlaw wants most to assume the mantle of "the greatest gunslinger who ever lived," and to avoid being forever relegated to second-best status. If only one outlaw survives the present contest, he will of course be remembered as the greatest. On the

FIGURE 15    **Payoff matrix for the Mexican Standoff**

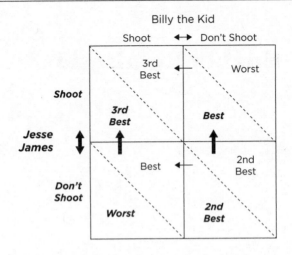

other hand, if both die or both survive, both will continue to be judged as coequal in their dastardly exploits.

Figure 15 summarizes these considerations in the form of a payoff matrix. Note that each outlaw has a dominant strategy to shoot. Consider Jesse James. If Billy shoots, Jesse prefers to shoot as well, to ensure that he doesn't fall to second place in the Gunslinger Hall of Fame. On the other hand, if Billy does not shoot, Jesse still prefers to shoot, though now to take first place. Of course, both are worse off when both shoot (tied and dead) than when neither shoots (tied and alive). So, the Mexican Standoff is a Prisoners' Dilemma.

Each outlaw has a dominant strategy to shoot, but neither pulls the trigger right away. Why not? Each knows that his own shot will not instantly kill and that, in fact, the other gunslinger will reflexively pull his own trigger in the moment of impact. Since shooting the other guy is tantamount to pulling the trigger on yourself, each side refrains from starting a fight. Both sides are safe . . . at least for a while.

The Mexican Standoff is an example of what's known as the Mutually Assured Destruction (MAD) Game. The crucial feature of MAD Games is that both sides can inflict devastating harm, even after suffering a devastating attack. The most famous MAD Game was played by the United States and the Soviet Union during the Cold War.

# Mutually Assured Destruction (MAD)

> There is a difference between a balance of terror in which
> either side can obliterate the other and one in which both
> sides can do it no matter who strikes first.
>
> —*Thomas Schelling,*\* The Strategy of Conflict *(1960)*

In March 1944, more than a year before the first ever detonation of an atomic bomb, *Astounding Science Fiction* published a short story called "Deadline," describing the inner workings of an atomic bomb in great and accurate detail.[3] The drama of the tale reflected the fear, widely shared among scientists at the time, that an atomic explosion might set off an uncontrollable chain reaction of destruction. *Astounding*'s publisher, John Campbell,[4] came up with the idea for the story himself, which he explained in a letter to the author (Cleve Cartmill) whom he had recruited to do the actual writing:

> They're afraid that [the atomic] explosion of energy would be so incomparably violent... that surrounding matter would be set off.... And that would be serious. That would blow an island, or hunk of a continent, right off the planet. It would shake the whole

---

\*Schelling won the 2005 Nobel Prize in Economics (along with Robert Aumann) for pioneering work that "enhanced our understanding of conflict and cooperation through game-theory analysis."

Earth, cause earthquakes of intensity sufficient to do damage
on the other side of the planet, and utterly destroy everything
within [thousands of] miles of the site of the explosion.

The story, as Campbell envisioned it, would be "the adventure of the
secret agent who was assigned to save the day—to destroy that bomb."
Sounds exciting, but *Astounding*'s readership wasn't particularly
impressed. In a reader poll, "Deadline" came in dead last among the
six stories in the March 1944 issue. Little did those readers know at
the time that the Bomb was more than just science fiction.

While scientists' fears of atomic chain reaction turned out to be
unfounded—the white sands of New Mexico didn't detonate[5]—the
first nuclear explosion, on July 16, 1945, did set off a *strategic* chain
reaction with its own scary implications. Just four years later, on
August 29, 1949, the Soviet Union detonated its own first nuke. Within
another four years, both sides had detonated fusion bombs, so-called
thermonuclear weapons a thousand times more powerful than the
first-generation bombs that decimated Hiroshima and Nagasaki. By
1960, the United States had stockpiled over 20,000 nuclear weapons,
while the Soviets had about 1,600 of their own, enough to blow each
other to pieces many times over.[6]

Both the United States and the Soviet Union felt compelled to keep
such large stockpiles to deter the other side from attacking first. After
all, even the most successful preemptive nuclear strike could not take
out all of the other side's missiles, many of which were buried deep
underground in isolated bunkers or carried on a hidden fleet of submarines. And even a small counterstrike of (say) a few dozen nukes
would inflict massive damage on the original attacker.

Some people may find this logic convincing and feel completely
reassured at the notion of maintaining peace by the threat of mutual
destruction. However, in fact, MAD is much scarier than that, and
everyone should be at least somewhat concerned that such schemes
could go horribly awry. Sure, it makes sense that no one ever

launched a nuclear attack during the Cold War. But can we really be confident that, should a similar situation arise in the future, we will be equally safe?

---

*Mutually Assured Destruction (MAD) has all the classic ingredients of a dangerous theory: (i) simple and compelling logic but (ii) hidden assumptions on which that logic rests, without which the theory could fail in dramatic fashion.*

---

There is always danger when we rely on an abstract theory to make predictions about the real world. We've encountered this "danger of theories" before, in the introduction, when we saw how traders' blind faith in the Black–Scholes formula led to the collapse of Long-Term Capital Management in 1998. That financial crisis was bad enough, but obviously, the stakes are much higher when we're talking about nuclear war. And let's be clear. Mutually Assured Destruction has all the classic ingredients of a dangerous theory: (i) simple and compelling logic but (ii) hidden assumptions on which that logic rests, without which the theory could fail in dramatic fashion. In particular, the MAD theory hinges critically on at least three key assumptions, without which the theory's prediction that there will be no war could turn out to be horribly wrong. These assumptions all relate to the effectiveness and credibility of each side's threat to retaliate with a nuclear counterstrike:

1. *Sanity*: Each side prefers to avoid being nuked itself.
2. *Capability to retaliate*: Each side is *able* to destroy the other side after a preemptive strike.
3. *Incentive to retaliate*: Each side is *willing* to destroy the other side after a preemptive strike.

The importance of sanity is fairly obvious. If either side actually wanted to be nuked, or to end life on earth as we know it, one easy way to do so would be to launch missiles at the other side. Hollywood screenwriters have had plenty of fun with this notion, concocting all sorts of reasons why someone might want to provoke a nuclear war, from insane generals (*Dr. Strangelove*) and internal rebellion (*Crimson Tide*) to intelligent machines indifferent to human suffering (*WarGames*) or actively seeking to take over (*Terminator*). Of course, all these imaginary movie scenarios are sufficiently far-fetched that most people might be left thinking that we are safer than we really are. The most realistic (though still hopefully low-probability) risk of MAD failure comes from the possibility that other assumptions, related to credible retaliation, might fail.

From the beginning, Cold War planners understood the importance of being capable of quick retaliation against any attack. That's why, since the days of Dwight Eisenhower, every US president has been constantly accompanied by an officer carrying a modified briefcase, known simply as "the Football," that contains the authentication codes and communications capabilities that the president needs to order a nuclear strike. Should the president send such an order, all it takes to launch is confirmation by the secretary of defense. Congress is not consulted, nor is anyone else.

This is quite unsettling—that two people can decide the fate of the world. However, more oversight would only slow down the process of retaliation, perhaps even giving other nuclear powers reason to doubt whether we are capable of launching an effective counterstrike. Such doubt could be fatal since, as I argue below in "A Dangerous President," any failure of certain retaliation could actually spark a nuclear war, even in "harmless" situations in which no nuclear power wants to harm any other.

## Example: Reagan's "Star Wars" Program

In March 1983, President Ronald Reagan announced the Strategic Defense Initiative (SDI). The purpose of SDI was to spur research so that the United States could—at some later date—launch ground-

and space-based systems to protect America from Soviet nuclear attack. Whether this so-called "Star Wars" missile defense system would have made the world a safer place is debatable, but one thing is certain: Reagan's public announcement that the United States sought to develop such a system instantly made the world a more uncertain place.

The Soviets reacted to Reagan's announcement with disbelief and suspicion. According to a US Naval Postgraduate School PhD thesis on the Soviet response:

> In 1984, the Soviet Committee for Peace and Against the Nuclear Threat [issued] a highly detailed and technical report, [citing] the SDI's probable enormous cost and extreme vulnerability to countermeasures as two reasons for concluding: "The assertions coming from the Reagan Administration that the new antimissile defense systems spell salvation from nuclear missiles for mankind are perhaps the greatest ever deceptions of our time."[7]

We're lucky that the Soviets didn't believe we were capable of building an effective nuclear shield. If SDI had been even potentially technically feasible at the time, the Soviets would have reacted with immediate alarm, no doubt ratcheting up their military spending to develop new weapons that could neutralize or evade the "Star Wars" threat. We could have been thrown into an entirely new space arms race, with who knows what unforeseen consequences and dangers.

## Example: A Dangerous President

Reagan's "Star Wars" missile shield has received plenty of attention over the years, for its potential to disrupt the world's MAD peace, but other potential dangers are more political in nature. During the Cold War, for instance, the Soviet Union could have been reasonably concerned that Americans might elect a president who would launch a preemptive nuclear strike. That may sound preposterous, but let me explain.

Suppose the American presidency were a dictatorship, i.e., a presidency for life. Even the most bloodthirsty, Soviet-hating American dictator would not dare to launch a preemptive strike on the Soviet Union, knowing the certainty of a devastating retaliatory blow. However, America is not a dictatorship. Indeed, our political system has the strange feature that the American president retains power throughout a "lame duck" period, during which time the current president knows who the next president will be. This creates a danger, as successive presidents can have very different views on foreign policy, including whether to launch a retaliatory strike. To see why such inconsistency can be dangerous, consider the following fictional scenario.[8]

Imagine that, in the waning days of the Cold War, America has elected a hard-liner (President #1) who feels certain that, absent the credible threat of nuclear retaliation, the Soviet Union would launch a preemptive strike against the United States. Imagine further that this president loses reelection to an opposing candidate (President #2) who takes a much more conciliatory approach. Indeed, President #2 believes that President #1's bellicose attitude is the real problem, that the Soviet Union would never attack and, even if it did, launching a counterattack would be the ultimate folly, as it would ensure the likely extinction of the human race.

What would you do, if you were President #1, the day after you lost the election? Remember: you believe that, once President #2 takes over, the Soviets will exploit his weakness and immediately launch a devastating preemptive strike on the United States. Obviously, you can't let that happen. You simply cannot let President #2 take power. You and your like-minded secretary of defense call in the military to explore the possibility of a coup to extend your presidency, but they refuse to violate their oath to support and defend the Constitution, despite your insistence that the future of the world hangs in the balance.

Restless and distressed, you turn to the only option that remains. You and the secretary of defense say a final prayer and launch a preemptive strike yourselves. That way, at least you can take out a goodly portion of the Soviet arsenal, leaving them with a smaller stockpile with which to decimate the United States. The way you figure it, at least a few Americans will survive, you among them, and the Soviets will be too busy trying to survive themselves to invade and finish us off.

Of course, the Soviet Union isn't a passive player in all this. Given the transparency and openness of American politics, they most likely know President #1's mind. Rightly fearing a preemptive strike during the interregnum between Election Day and President #2's swearing in, the Soviets might "rationally" launch a preemptive strike of their own. In the end, humanity might not even survive election night.

Mutually Assured Destruction Games are so-called "dynamic games." The defining feature of dynamic games is that they take place in real time, with each player capable of quickly observing and reacting to changes in the other's behavior. In the Cold War context, both the United States and the Soviet Union could observe and respond to any nuclear strike since (i) once launched, missiles are visible to radar, and (ii) no preemptive attack could eliminate all missiles, especially those carried by each side's fleet of nuclear-attack submarines. In the Mexican Standoff, similarly, each outlaw could pull his own trigger after being hit himself.

---

*A game has "dynamic moves" when it takes place in real time, with each player capable of quickly observing and reacting to changes in the other's behavior.*

---

Another key feature of MAD Games is that they are "waiting games," continuing in a fixed status quo until someone finally "pulls the trigger." Not all dynamic games are like that. For instance, in the Dynamic Pricing Game considered next, each airline is capable of reversing its pricing move. This means that "mistakes" in dynamic pricing can be undone at little or no cost, whereas mistakes in MAD lead to assured mutual destruction.

## Example: Dynamic Pricing*

Imagine that only two airlines (say, Delta and American) operate direct flights on some route (say Chicago–Atlanta). Imagine further that customers interested in flying this route first check what prices the airlines are charging and then book with whichever airline is cheaper. For simplicity, assume that customers only check prices once and that customers don't have any loyalty or preference between the airlines. Moreover, suppose that every traveler is willing to pay $200 for a ticket, but that it only costs $100 to provide this service.[9]

If Delta and American could form a cartel to maximize their collective profit, they would each set a price of $200 to capture every traveler's full willingness to pay for the flight. As competitors, however, Delta and American have an incentive to undercut one another on price. Yet whether they will actually offer lower prices depends on the details of the game they play. Consider two alternative options:

1. *Auction*: Each traveler requests secret price quotes from the airlines, who are not allowed to communicate with one another, and selects whichever airline has the lower price. (If the airlines offer the same price, the traveler flips a coin to decide.)
2. *Posted Prices*: The airlines maintain public prices that travelers and the airlines can see. Each traveler views these prices and

---

*For more on the real game that inspired this example, see Game-Changer File 1, "Price Comparison Sites."

**FIGURE 16**  Average per-traveler profit for airlines in the auction scenario

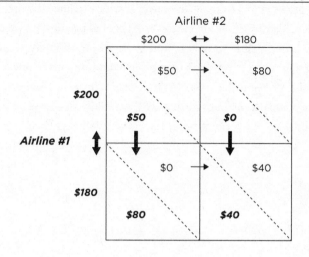

selects whichever airline has the lower price. (If the airlines have posted the same price, the traveler flips a coin to decide.)

Consider first the auction. Figure 16 illustrates each airline's average per-traveler profit, depending on how they set prices (assuming, for simplicity of the exposition, that only prices $200 and $180 are possible). For instance, if one airline charges $200 while the other charges $180, the traveler is sure to go to the lower-priced airline, giving that airline profit of $180 – $100 = $80. On the other hand, if both airlines charge $180, whether each airline makes a sale depends on the traveler's coin flip. So, each airline makes $180 – $100 = $80 half of the time, and nothing the other half of the time, for an average profit of $40. (Similarly, when both airlines set a price of $200, they each make an average profit of $50.) Note that each airline has a dominant strategy to price low, but when both price low, both are worse off than when both price high. So, this is a Prisoners' Dilemma.

Unfortunately, since the airlines offer secret price quotes in the auction, there is no way for them to use the threat of retaliation to escape this Prisoners' Dilemma.[10] Not so with posted prices. Since travelers can check prices at any time, no airline ever wants to have the highest posted price. Should either airline lower its posted price, then, we can expect the other airline to match, and to do so immediately. Consequently, any move to lower one's own price is tantamount to lowering both airlines' prices in lockstep.[11] With nothing to gain from undercutting the competition, both airlines will naturally sit content with a posted price set at $200, exactly *as if* they were a collusive cartel.

# CHAPTER 5

# Build Trust

The man of upright life is obeyed before he speaks.

—*Confucius*

You've just received an email from the treasurer of the Nigerian National Petroleum Company, who needs your help to transfer a windfall of $40 million out of the country. He will pay you $4 million for your trouble, but first you need to send $10,000 to open a Nigerian bank account. Sounds like a pretty good return on investment, right? Of course not! This is the infamous "Nigerian scam," one that you've probably seen and resisted yourself. But plenty of people fall for it. Even as far back as 1997, the Secret Service Financial Crimes Division reported "confirmed losses just in the United States of over $100 million in the last 15 months" from the scam.[1]

You might never fall for the Nigerian scam, but what about this? After searching the online price comparison site PriceGrabber .com for a new camera, you notice that BPPhoto.com is selling the Canon EOS Rebel T3 for only $409, a steal relative to the $499 price charged by Amazon and other featured sellers.* You've never heard of BPPhoto, but this is the sort of deal that doesn't come along every day.

---

*For more on PriceGrabber.com and this particular vignette, see Game-Changer File 1, "Price Comparison Sites."

All you need to do is pay the money and BPPhoto promises to send you the camera. Would you do it? If so, what's the difference between BPPhoto and that treasurer from Nigeria? Why trust one but not the other? This chapter digs into issues of trust, focusing primarily on two questions:

1. What can a trustworthy player achieve that an untrustworthy player cannot?
2. How can trust be earned?

We live in a mostly lawful and ethical society, where people treat others honorably and well, just for the sake of doing so. Within that context, trust is given and received with grace. If that were the end of the story, this would be a very short chapter. However, even the most shining city on a hill will always have some residents who don't care what's right or wrong, who don't care about anyone else's well-being. Psychiatrists have a name for these people: sociopaths, also known as psychopaths.[2]

The term "sociopath" brings to mind movies like *Psycho* and stories of crazy serial killers, but most sociopaths actually live normal lives like you and me, with one key difference: sociopaths don't give a damn about anything or anybody but themselves. No one knows exactly what underlies sociopathy, or exactly how many sociopaths are among us; estimates vary from 1 to 4 percent of the population. That's common enough that, in all likelihood, someone at your workplace or in your social circle is a sociopath. Such people are more likely to betray you or your organization for their own personal gain. It's important, then, that all your "personnel decisions"—from whom you date to whom you hire or promote—take into account the possibility of sociopathy.[3]

My point here isn't to scare you, so that you see sociopaths in every shadow, but to emphasize that trust is never automatic. We must always decide whether to trust, knowing that there is a possibility that we might be betrayed. Fortunately, the deeper message of game theory is that trust is possible even in a completely selfish world. Furthermore, trust can powerfully transform our lives as it opens up strategic possibilities for all sorts of win-win outcomes that otherwise would be closed to us.

## Trust in a Selfish World

Being a criminal isn't easy. Constantly on the lookout for law enforcement, you've also got to worry that your partners in crime may turn on you. Unlike businesspeople in the legal world, most criminals can't sign binding contracts or settle disputes in a court of law. So, cheaters really do prosper. Making matters worse, sociopaths are much more common among the criminal class than in society at large. (According to a 2002 estimate published in *The Lancet*, 47 percent of male prisoners and 21 percent of female prisoners are sociopathic.)[4] Thus, much more so than in the legal world, trust must be earned in the criminal world. This raises strategic challenges unique to criminal enterprise. However, the way in which criminals overcome these obstacles is instructive for us all, as the same methods can be used to strengthen trust in other extralegal relationships, with everyone from family members to office gossip-mates and romantic partners.

Consider the strategic problem faced by two criminals who would like to do a deal to trade 50 kilos of cocaine for $1 million, in the Coke Deal Game whose payoff matrix is shown in figure 17.[5] The buyer knows that, once he pays the money, the seller will have an incentive just to take it, walk away, and sell the drugs to someone else. Indeed, whether the buyer pays or not, the seller has a dominant strategy not to deliver the drugs. Similarly, the seller knows that once he delivers the drugs, the buyer will have an incentive just to take them and pay nothing. Indeed, whether the seller delivers the drugs or not, the buyer has a dominant strategy not to pay. As long as both buyer and seller play their dominant strategies, however, no deal gets done and both are worse off than if they had made the trade. So, the Coke Deal Game is a Prisoners' Dilemma.

**FIGURE 17**   Payoff matrix for the Coke Deal Game

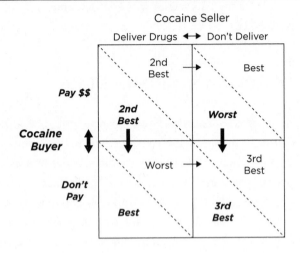

Real-world criminals do plenty of deals, so they must have ways to overcome this incentive to cheat one another. How do they do it? First and most obviously, criminals have internalized the lessons of cartelization and self-regulation from chapters 2 and 3, forming huge umbrella organizations and incentivizing good behavior among their "associates" with the threat of murder (or worse) for anyone who gets out of line. Moreover, even the most sociopathic criminal recognizes the benefits of (i) cultivating a reputation for being a trustworthy business partner and (ii) building relationships for repeat business, just as in the legal business world.

## What If Only One Side Can Be Trusted?

It's easy to see how two criminals can get a deal done if both are concerned about maintaining a good reputation, and hence both can be trusted not to cheat. But what if only one side to a deal is trustworthy? Fortunately for criminals, it's possible to escape the Prisoners' Dilemma of illegal transactions as long as either side can be trusted, since that player can leverage its reputation to make a credible promise. For instance, a trustworthy cocaine seller can promise this to the buyer: "First, give me the money. I promise that I will then give you the drugs."

Put yourself in the shoes of the cocaine buyer, pondering this promise. If you don't give the money, you definitely don't get any drugs. If you do give the money, however, you will get the drugs as long as the seller honors his promise. And, as long as the seller's reputation is worth more to him than the value of this particular deal, you have reason to trust the seller to follow through. So, your choice is really between the two outcomes whose payoffs are boxed in figure 18: "Don't Pay + Don't Get the Drugs" or "Pay + Get the Drugs." Given this choice, you'll pay the money. Of course, being a criminal yourself, you'd love to cheat the seller and get the drugs for nothing. However, by design, the seller's promise forces you to move first, so it's not possible for you to cheat.

FIGURE 18   The buyer's choice after the seller's promise

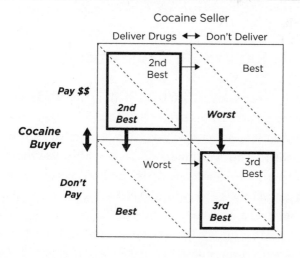

---

*Players can escape the Prisoners' Dilemma as long as either of*
*them can credibly make the following promise:*
*"First, you refuse to confess. I promise that I will then also*
*refuse to confess."*

---

There are two key ingredients in any promise, related to observability and credibility:

1.  Some player (the "last-mover") can observe and respond to the other's move.
2.  The last-mover can credibly commit to how it will respond.

Of course, the trick is how to credibly commit to a promise that you will ultimately prefer to break. There are many ways to build up one's credibility, but the best is also the simplest: conduct yourself at all times with the utmost integrity. That way, others will simply trust your word without a second thought.

---

*The best way to build up your credibility is to conduct yourself*
*at all times with the utmost integrity.*

---

## It Pays to Be Trustworthy

The fact that only one side needs to be trusted has important implications for how commercial transactions (legal and otherwise) are organized. Indeed, as long as counter-parties to a trade have any doubt about each other's trustworthiness, there is a profitable niche for trusted third parties to coordinate deals. First, trusted intermediaries can provide and charge for information that reduces uncertainty about the quality of a deal. For instance, Carfax reports help assure buyers that a used car for sale has never been totaled, flooded, or sent back to the manufacturer as defective under state "lemon laws." Once there is less uncertainty, there is less to lie about and less need for trust to get the deal done. Second, trusted parties can leverage others' trust in them to earn an extra profit. Consider how the used-car industry "certifies" used cars to limit buyers' uncertainty about their quality. Certified used cars can then be sold at a premium.

## Example: Certified Pre-owned Vehicles

> [Purchasing a used car from] a private seller requires faith that the owner is being honest about the car's history. Buying from a used-car dealer necessitates trust of a whole different order, namely, that the dealer knows anything about the car's past. . . . Another avenue has become a popular alternative: purchasing a used car from a manufacturer's certified program.
>
> —Car and Driver, *March 2009*[6]

The 24 Hours of Le Mans is perhaps the most eccentric race in all of automotive sport. Held annually since 1923, drivers race continuously for twenty-four hours over a 13.6-kilometer circuit that includes stretches of dedicated track, as well as well-used roads through the city of Le Mans, France. Unlike Formula One, which rewards aerodynamics and acceleration, endurance races like Le Mans test the whole vehicle and its ability to survive twenty-four hours of rough duty. As you can imagine, durability was an especially important issue in the early days of the race, when automobile makers were still learning how to build reliable cars. In the 1920s, Le Mans was dominated by Bugatti, Bentley, and Alfa Romeo. But then a bold new competitor burst onto the scene: Aston Martin. In 1928, Aston Martin entered its first vehicle in the 24 Hours of Le Mans; by 1933, it had swept the field, taking all the podium places in its class.[7] No wonder that the Aston Martin is James Bond's favorite ride.

For all its undeniable coolness, the Aston Martin has one serious drawback: depreciation. In August 2011, *Popular Mechanics* magazine published a list of "10 Cars That Depreciate Like a Stock Market Crash," including the Aston Martin. This wasn't a scientific study, mind you, but rather just a list of fancy used cars that had recently sold at prices dramatically below their original sticker price.[8] Nonetheless, it seems remarkable that a 2001 Aston Martin DB7 Vantage V12 coupe with only 32,000 miles, originally priced between $160,000 and $200,000, recently sold for just $32,100 on eBay.

Why do high-performance cars like the Aston Martin depreciate so much in value? One obvious reason is that the value of such vehicles depends on how well they have been maintained. And it's natural to suspect that anyone willing to part with such a beautiful automobile may not have done enough to take care of it properly. To solve this problem, Aston Martin launched a certification program for preowned cars called "Aston Martin Assured" in 2009 to provide, in the words of business development director Philipp Grosse Kleimann, "complete and unrivaled peace of mind and security for our customers."[9] When I checked in October 2012, Aston Martin's website listed a few dozen Assured vehicles available in the United States, the cheapest being a 2007 Vantage V8 coupe for $69,995.[10]

Mercedes-Benz was the first manufacturer to offer such a certified pre-owned (CPO) vehicle program, in 1989, followed by Porsche in 1991 and Lexus in 1993. Early on, certified pre-owned cars were mainly low-mileage vehicles coming off a lease and were seen as an alternative to a new car. As explained by *Consumer Reports'* Rob Gentile: "These were two- to three-year-old cars with less than 50,000 miles on them.... Now we have five- to eight-year-old cars with more than 60,000 miles on them [and most manufacturers offer CPO vehicles]."[11] Indeed, Cars.com estimates that "1.6 million of the 17 million used cars sold by dealers each year are factory-certified."[12]

Car dealers and manufacturers benefit by certifying used cars, for several reasons. First, used-car buyers who are satisfied with their vehicles are more likely to buy a new car of the same make. As Jerry Cizek, former president of the Chicago Automobile Trade Association, told Cars.com: "One reason for offering certified pre-owned cars is that it's a way for automakers to keep you in their family. If they attract you into a pre-owned car and you're happy, you eventually will buy the same brand when you purchase new, and they are hoping for repeat new-car sales."

Used cars whose quality has been certified also sell for more, at a premium ($800 to $1,300) that is typically much more than the dealer's cost of conducting the certification inspection. This allows dealers to make a profit on certified cars, but raises the question why buyers don't just hire their own mechanic to conduct an inspection. As Tom Kontos, vice president at ADESA, a vehicle remarketing firm, explained to Cars.com:

> If you have gone over the vehicle thoroughly, or hired a mechanic to help inspect it, and combine that with an extended warranty, you could create a quasi-certified vehicle.... That may be a better way to go for people who have the ability and time to go through that themselves and save some money.

Of course, not everyone has that kind of time on their hands.[13]

Offering a certification program also allows dealers to establish a reputation for selling only high-quality vehicles. Since a problem vehicle will not be certified, buyers will naturally interpret the absence of certification as evidence of low quality and only buy such cars at low prices. This changes dealers' incentives to acquire problem cars in the first place, knowing that they won't be able to unload them as profitably as otherwise. Understanding this, buyers can visit a dealer who offers a certification program, trusting that the vehicles on the lot have been more carefully screened and selected.

## Building Trust

Trust must be earned. But what if you have not been trustworthy in the past? Fortunately, it's never too late to turn the page and become a trustworthy person. That said, you must also be *trusted*, and whether others are willing to trust you is not entirely in your control. First, others must really believe that you have changed. But how can you make them believe, given your past misdeeds? To learn whether you are trustworthy, they first need to trust you one more time. Fearing another betrayal, however, even your most loved ones may refuse to give you another chance. To earn that chance, you may need to do something dramatic and unexpected, to signal that something really has changed.

### Example: Daddy's Dessert Dilemma*

> Like many a desperate parent, I have decided
> to resort to bribes.
>
> —*Heidi Grant Halvorson,* Succeed *(2010)*

I am blessed with three small children, but sometimes my little darlings can be a bit, how shall I say, challenging. Especially at dinnertime. I serve vegetables with dinner each evening and, after the kids finish eating, I decide whether to serve dessert. I want my kids to have the healthiest possible meal. Bearing that in mind, my top priority is to get the kids to eat their veggies, while my secondary goal is not to serve dessert. As you can imagine, my children and I don't see eye to eye on this. Their top priority is to get dessert, while their secondary goal is not to eat their veggies. Our resulting payoffs are shown in figure 19.

Note that the kids have a dominant strategy not to eat their veggies, while I have a dominant strategy not to serve dessert. However, we all prefer the outcome with veggies and dessert over what happens (no veggies and no dessert) if we all play our dominant strategies. So, this game is a Prisoners' Dilemma.

---

*This story is a fictional account of a game that I might play if I were a single dad. In fact, my wife solved this dilemma long ago, in her own way, by cultivating our kids' taste for vegetables. Some of our favorite family foods now are asparagus, artichoke, broccoli, and roasted cauliflower.

**FIGURE 19** Payoff matrix for Daddy's Dessert Dilemma

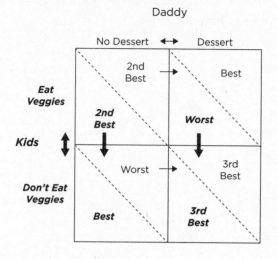

How can we escape this Prisoners' Dilemma? Most parents know the answer, instinctively, whether or not they know anything about game theory. To get the kids to eat their veggies, I just need to make the following promise: "If you eat your veggies, I *promise* to serve dessert." As long as the kids believe this promise, their choice is between (i) eating veggies and getting dessert or (ii) not eating veggies and not getting dessert. Since getting dessert is their top priority, they will eat their veggies, making me happier as well. See figure 20 where, reflecting the promise that I have made, my option not to serve dessert after the kids eat their veggies has been crossed out. For this to work, however, the kids have to trust me to follow through on my promise.

As a game theorist, I have been cultivating my children's trust since the day they were born. We even have a code phrase ("On your honor, Daddy?") that the kids can use when they want to double-check if I'm telling the truth. I will never *ever* lose this "honor" before my children, since I understand how important trust will be to change games for the better, throughout our lives together. But what if I hadn't already earned my children's trust? What if, actu-

**FIGURE 20**  Daddy's Dessert Dilemma *if* Daddy promises dessert after veggies. The kids decide between outcomes where their payoffs are circled.

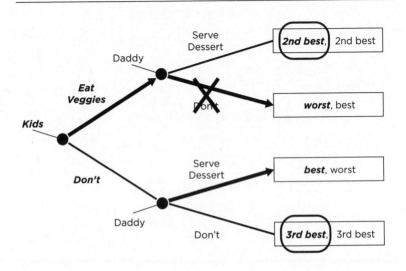

ally, I had taught them not to trust me by routinely deceiving them to get them to do what I want?

Unfortunately, without already-established trust, promises are worthless. To see why, suppose that I have a history of being an untrustworthy daddy, but that I've decided to make a change and truly promise to give the kids dessert if they eat their veggies. The kids hear this promise but know that, once they eat the veggies, I will have an incentive to renege on the deal and withhold dessert. Anticipating this, the kids may never even give me "just one chance" to prove that my promise is legitimate. And as long as I am unable to prove that I will follow through on my promise, they will never come to trust me enough to give me that chance.

My only hope of earning their trust is to do something that shakes up their firmly held notion of my untrustworthiness. Fortunately, there are many ways to build trust, even in relationships with a history of failed promises. For example, I might call a family meeting and make the following little speech:

> Guys, we're going to make some changes that I think will make us all happier and healthier. I used to never serve dessert, no matter what, but now I'm willing to let you have dessert if you eat your veggies. I know you don't believe me, because I changed my mind and didn't give you dessert that other time when I said I would. But it's going to be different this time. To prove it to you, I'm going to let you have dessert "for free" all this week. If I just wanted to trick you into eating veggies, I would never do that. So you can trust me when I say that, starting next week, I *promise* to serve dessert if you eat your veggies. I'm going to keep that promise, every time, and let me tell you why. Because I'm your daddy. And I need you to be able to trust me, not just with dessert but with everything.

# CHAPTER 6

# Leverage Relationships

*The great enforcer of morality in commerce is the
continuing relationship, the belief that one will have to do
business again with this customer, or this supplier.*

—*Martin Mayer,* The Bankers *(1974)*

Your car has started to have some trouble, just a few weeks
before you are set to drive off to a new home across the country. You take it to a local mechanic and, while explaining the
situation, let slip the fact that you are about to leave town for good.
Later, while you're waiting for the mechanic to complete his inspection,
a disturbing thought pops into your head. Was it a mistake to tell him
that you are moving? After all, the mechanic now knows that there's
no prospect of repeat business or positive word of mouth from you,
regardless of how good a job he does. Will he charge you more or, worse
yet, not be as careful when inspecting and repairing your vehicle?

Henry Schneider, an economics PhD student at Yale, came up with
a clever way to find out.[1] Following an undercover methodology developed by the Automobile Protection Association, a Canadian consumer
protection group, Schneider drove his 1992 Subaru Legacy L Wagon
to forty different repair shops in the New Haven area during the summer of 2005. He and his car were exactly the same during every visit,
with one exception: at half of the shops, Schneider claimed to be moving from Connecticut to Chicago, while at the other half, he claimed
to be about to drive round-trip to Montreal. (The one-way distance to

Chicago is about the same as the round-trip distance to Montreal and back.) At each shop, Schneider had a thorough inspection performed and made note of the mechanic's recommendations on what needed to be fixed. Then, begging off with the explanation that he would "think about it," Schneider repeated the whole process at the next place.

The car's repair needs were the same at every stop. A loose battery cable causing intermittent battery failure provided a credible excuse for bringing the car in, but there were also more serious problems: (i) low coolant, essential to fix to avoid engine overheating; (ii) a missing taillight; (iii) worn spark plug wires, including one wire that did not fit properly into the engine block and allowed debris and rainwater to enter and corrode the cylinder head; and (iv) an exhaust pipe leak beneath the driver's seat that allowed fumes to enter the car through an open window.

Overall, the mechanics' performance was unimpressive. Only four of the forty mechanics who inspected the vehicle identified three or more of the four important issues. Indeed, the missing taillight was found in only five visits, while the low coolant level—easily visible from the engine compartment—was discovered in only eleven visits. The "good news," though, is that the mechanics diagnosed about the same number of problems, and recommended about the same number of unnecessary repairs, no matter whether Schneider claimed to be leaving town.

The bad news is that they charged Schneider much more on average for the inspection when he claimed to be leaving town for good ($59.75) than when he said he would be returning ($37.70). Since Schneider randomly decided what to tell each mechanic and made sure to keep everything else exactly the same (even his clothing, always khakis and a polo shirt), this effect must have been caused somehow by the fact that he told them he was moving.

Some people may look at this and jump to the conclusion that mechanics must be bad people, ripping off customers who are about to leave town. In fact, the most natural interpretation is much more benign: that mechanics offer discounts to people they hope will return for more work in the future. Indeed, inspections might even be a loss leader in the auto-repair business, as mechanics invest to get to know both you and your car. If you're leaving town, mechanics might naturally feel that it's fair to charge you a price that more closely reflects their true cost of inspecting the vehicle. (This "mechanics aren't bad guys" interpretation is bolstered by Schneider's finding that they recommended about the same number of other repairs, and charged about the same for those repairs, regardless of whether he claimed to be leaving town.) The story here is therefore about how reputation and repeat business induce mechanics to charge *less* for an inspection than they otherwise would.

Mechanics, dentists, and restaurants are just a few of the many types of businesses that rely on repeat customers. The prospect of repeated interaction is a powerful inducement for players to exert extra effort and cooperate in various ways. That said, repetition by itself is not enough to induce cooperation. To see why, consider a "repeated Prisoners' Dilemma" in which the same two players play a Prisoners' Dilemma game many times, and care about the total payoff that they earn over the life of the relationship. As long as the other player is always going to confess, your best response is always to confess as well. For this reason, everyone always confessing is an equilibrium of the repeated Prisoners' Dilemma, just as it is in the one-shot Prisoners' Dilemma.

This failure to help one another may seem like bad news, but some relationships require the *threat* of future breakdown to sustain cooperation today. This is especially true in the repeated Prisoners' Dilemma, where players' incentives to cheat one another must be continually held in check. Indeed, in order for players in the repeated Prisoners' Dilemma to cooperate, it's essential that any cheating trigger a "punishment phase" during which cooperation ceases. Such temporary relationship breakdown is painful to both players, including the "victim" who was cheated, but essential to incentivize good behavior in the first place.

What if the temptation to cheat is just too great to be deterred by any credible punishment? In that case, players can still maintain a cooperative relationship by scaling down the stakes. For instance, consider the Coke Deal Game described in chapter 5, in which two criminals are trying to do a drug deal, and suppose that an especially large shipment of cocaine arrives one day. There is just too much money at stake for the buyer and seller to credibly commit to keep their end of any deal in which all of the drugs are traded. Fortunately for the criminals, the solution is simple: just split the shipment up into several smaller batches. As long as each individual transaction is small enough, no one will want to spoil the rest of the deal by cheating on any single batch.

**REPUTATION WITHOUT REPEAT BUSINESS**

What if you only interact once with someone, but need their trust to get a deal done? Fortunately, all the strategic benefits of repeat business can still be had if there is some way for others to verify how you have handled yourself in previous deals. For instance, online marketplaces such as eBay have introduced "reputation scores" to help address the problems of anonymity and non-repeat business over the Internet. By design, reputation scores transform anonymous eBay transactions into (at least somewhat) observable encounters that others can use when deciding with whom to do a deal. Thus, sellers know that how they treat each customer will affect not only whether that specific customer returns for repeat business, but whether others will choose them as well. In this way, eBay reputation harnesses the power of relationships to create a relatively safe environment in which to buy and sell on the Internet.*

## Disrupting Harmful Cooperation

Cooperation among one group can sometimes hurt society at large. For example, business cartels collude to raise prices, helping themselves but hurting consumers. Antitrust authorities discourage such

---

*eBay has done much to create a trustworthy trading platform, but there is still room for improvement. See Game-Changer File 5, "eBay Reputation."

harmful cooperation not only by actively seeking to catch colluders, but also by changing the game to undermine colluding businesses' ability to evade detection.

Among all the game-changing weapons in the antitrust arsenal, the most powerful is actually *letting some guilty firms go free*. In 1978, the US Department of Justice (DOJ) introduced a first-of-its-kind Corporate Leniency Program through which firms could qualify for amnesty by being the first to confess to illegal cartel activities. For the next fifteen years, however, DOJ didn't crack a single international cartel or major domestic cartel due to information gleaned from amnesty applications. The problem was that DOJ retained some prosecutorial discretion regarding whether to grant amnesty to confessing firms. Faced with the risk that they might still be prosecuted even after confessing, firms only had an incentive to confess if they were fairly certain that others were also likely to confess. But as everyone felt shy about confessing, no one had much reason to worry. The natural equilibrium of this game was for all cartel members to stay quiet.

That all changed in 1993, when DOJ revised its Corporate Leniency Program to *guarantee* complete amnesty to the first confessing firm, regardless of the seriousness of their crimes. This guarantee made immediate confession the only risk-free strategy for firms, and the floodgates opened. Deputy Assistant Attorney General Scott Hammond, head of criminal enforcement at the DOJ's Antitrust Division, explained the dramatic effect that leniency had on cartel detection and antitrust enforcement:

> The advent of leniency programs has completely transformed the way competition authorities around the world detect, investigate, and deter cartels. Cartels by their nature are secretive and, therefore, hard to detect. Leniency programs provide enforcers with an investigative tool to uncover cartels that may have otherwise gone undetected and continued to harm consumers. While the notion of letting hard-core cartel participants escape punishment was initially unsettling to many prosecutors, the Anti-

trust Division recognized that the grant of full immunity was necessary to induce cartel participants to turn on each other and self-report. [As a result,] in the United States, companies have been fined more than \$5 billion for antitrust crimes since Fiscal Year 1996, with over 90 percent of this total tied to investigations assisted by leniency applicants.[2]

The genius of corporate leniency is that it can trigger an avalanche of confessions from the merest grain of doubt among cartel members about the stability of the cartel. Cartel members keep their communication to a bare minimum, to evade detection, but such radio silence creates room for misunderstandings which can bloom into concern and even outright panic. For instance, consider a cartel run by a group of mid-level pricing executives, kept secret not only from antitrust authorities but also from their own bosses (who forbid all illegal activity).

Suppose that some firm in the cartel starts setting lower prices than it is supposed to. Perhaps the pricing executive at that firm is getting a bit greedy, and just needs a stern talking-to at the next cartel meeting. Or, more concerning, perhaps he has been discovered by his bosses and ordered to stop fixing prices. Such a development could trigger the breakup of the cartel but, absent corporate leniency, cartel members would still have little reason to fear being caught by the antitrust authorities. After all, the firm that has discovered the cartel's activity has every incentive to keep quiet about how one of its employees has been breaking the law. Once corporate leniency is part of the picture, however, firms that have uncovered wrongdoing within their own ranks have a strong incentive to confess, to secure amnesty.[3] Understanding this, everyone else in the cartel may also conclude that their only hope is to confess themselves—and to do so *first*—creating a veritable stampede to the DOJ.[4]

This idea that doubt can undermine cooperation applies much more broadly than just to cartels and antitrust enforcement. Indeed, the glue to any transactional relationship is "the belief that one will have to do business again with this customer, or this supplier." Take away that belief, or reduce the certainty with which it is held, and you'll weaken the trust that is at the heart of any successful relationship.

## Example: "Men of Honor"

Secret societies are part of the fabric of American life. Fourteen US presidents, from George Washington to Gerald Ford, are believed to have been Master Masons in the Freemasonry movement, a fraternal organization that guards the privacy of its meetings with special signs, secret handshakes, and passwords. Even as late as 2004, both leading candidates for president (George W. Bush and John Kerry) were members of the same secretive college fraternity, Yale's Skull and Bones.

Such societies may seem odd to outsiders, but they have a significant economic impact. Given their special bond, members of such groups have an automatic kinship and genuinely desire to help one another. This allows them to trust and recommend one another, and do business together, even if they are not personally acquainted. Of course, fraternal organizations like the Masons and Skull and Bones aren't really secret, nor do they want to be.

But there's another secret society, far more important and pervasive in America than any of these others, where secrecy is absolutely essential. Tightly knit like a family, this society admits only the most successful businessmen into its ranks, initiating them by old rites in which blood is shed, sacred icons are burned, and initiates swear their lives upon a solemn vow. Members conduct some business with nonmembers, when it's especially profitable to do so, but all core operations are conducted and coordinated in-house.

This society first took root in the 1890s in New York City, prospering and growing due to its members' loyalty and close cooperation, ultimately flowering in the 1920s into a powerful national (and even international) organization. Yet few in America even suspected its existence until the late 1940s. Who are these people? They call themselves "men of honor," but we know them as mafiosi and their organization as the American Mafia, or Cosa Nostra ("this thing of ours").

When a new member is initiated (or "made") into the Mafia family, he swears to abide by a variety of rules, such as "never discuss family business with nonmembers," "never kill another member without boss approval," "never commit adultery with a member's wife," and so on. But by far the most important element of the Mafioso code is *omertà*, absolute silence and noncooperation with law enforcement. Any member who violates this code will be tortured and killed. Not only that, his "family" will likely send his best friend to do the dirty work—all meant as a powerful warning and sign that family comes first.

Omertà is a highly effective shield against law enforcement's efforts to induce insiders to break ranks and inform on others in the organization. After all, the Prisoners' Dilemma isn't much of a dilemma if you expect a confession to result in your own murder. Indeed, despite numerous public hearings in the 1950s, no member even acknowledged the existence of the Mafia until 1963, when a low-level soldier in the Genovese family named Joseph Valachi testified to details of its operations. Valachi faced the death penalty for murdering a fellow inmate who he believed had been sent by the Genovese boss to assassinate him. All Mafia ties now cut, Valachi's only hope for survival was to cooperate and testify, after which he spent the rest of his life safely locked behind bars.

Valachi's testimony was a terrible blow to the Mafia, as politicians and law enforcement faced intense pressure to put a stop to Mafia operations.[5] Congress responded in 1970 with the Racketeer Influenced and Corrupt Organizations Act (RICO), which made mere membership in a criminal organization a punishable offense. That same year, Congress authorized the creation of the Witness Protection Program, as part of the Organized Crime Control Act, guaranteeing the safe and secret relocation of any federal witness and his family. RICO allowed law enforcement to arrest any known mafioso essentially at will, while witness protection took the teeth out of the Mafia's practice of punishing traitors. We might therefore have expected a flood of informers and wholesale dissolution of the entire Mafia network. But it didn't work out that way.

Dozens of Mafia figures were put behind bars, but no one significant took the bait of witness protection for over two decades. Obviously, the continuing benefits of remaining true to the Mafia family were more important to these mob bosses than avoiding prison.[6] Even so, RICO arrests were a disruption, forcing the Mafia to promote new bosses and recruit new soldiers more quickly and perhaps less carefully than before. It's easy to imagine how such internal turmoil could have destabilized the organization and even undermined some mobsters' sense of family loyalty. Whatever the ultimate cause, law enforcement finally got its big break in 1991, when the second-in-command of the Gambino family, Salvatore "Sammy the Bull" Gravano, broke ranks and testified against his boss, John Gotti.

Unlike any boss before him, Gotti actually courted publicity.[7] This bothered many of his underlings, who didn't like the extra "heat" that came with such attention. After Gotti and Gravano were arrested, the FBI played secret tapes to Gravano in which Gotti criticized him for "creating a family within a family" and wondered why everyone who partnered with Gravano kept winding up dead. The resulting breakdown of trust created the opening that the FBI needed and, in November 1991, Gravano became a government witness. Gravano's betrayal opened the floodgates, as key Mafia figures everywhere apparently lost faith in omertà and chose to testify.

Ironically, Gravano was eventually undone by the very "snitch" culture that he himself had helped to create. After leaving witness protection in 1995—even publicly taunting the New York Mafia: "They send a hit team down, I'll kill them.... Even if they get me, there will still be a lot of body bags going back to New York"—Gravano built up an ecstasy-trafficking ring grossing over $500,000 per week. But informants in that ring eventually helped authorities pin Gravano with a twenty-year sentence. He remains in federal prison today where, according to another jailed mobster's account, he only ventures out of his cell for food.[8]

# The Twin Pillars of Cooperation

The American Mafia took so long to crack because it was supported by both of what I call the Twin Pillars of cooperation:

1. *Intrinsic desire* to cooperate, especially if others are cooperating.
2. *Power to punish* others who fail to cooperate.

Players can support a cooperative relationship under either of these pillars. For instance, imagine a "true love" romance where each partner cannot imagine hurting the other. Such lovebirds may lack the power to punish, but that doesn't really matter since both want to cooperate. On the other hand, in a repeated Prisoners' Dilemma, neither player has an intrinsic desire to cooperate but each can credibly commit to "punish" any failure to cooperate by no longer cooperating itself. (More on this later.)

### "INTRINSIC DESIRE" TRUMPS THE "POWER TO PUNISH"
Either pillar can support cooperation, but "intrinsic desire" is the most secure foundation for a cooperative relationship. When players are intrinsically motivated to cooperate, it doesn't matter how long they expect their relationship to last. When cooperation is supported solely by the threat of punishment, however, each player must also believe (and believe that his/her partner believes) that the relationship is likely to last a long time. Why? Suppose that I believe our relationship is likely to end soon, regardless of what we do. I now have an incentive to betray you, since I know that you won't be able to punish me once we've gone our separate ways. That's what happened with Joseph Valachi, the first mob informant, after he decided that the Genovese family was trying to kill him. More subtly, if I believe that you believe that our relationship is coming to an end, I have an incentive to preemptively betray you, to avoid being betrayed myself.

That's what happened with Sammy Gravano, once the FBI convinced him that John Gotti planned to turn against him.

---

*"Intrinsic desire" is the most secure foundation for a cooperative relationship. Given just the "power to punish," cooperation can only be sustained if players believe it can be sustained for a long time to come.*

---

It took decades to crack collusion in the Mafia, because it was protected by both of the Twin Pillars of cooperation. Collusive arrangements that rely solely on the threat of punishment, on the other hand, are much more vulnerable to disruption. That said, combating collusion effectively requires a mind shift to focus the fight on the key strategic factors that facilitate collusion rather than on the symptoms of collusion. For instance, the medical profession has long battled biofilms (collusive bacterial communities) using conventional treatments that aren't very effective. Once medical professionals viewed the biofilm problem from a game-theory perspective, however, they hit on a novel and promising "Trojan Horse" approach to defeat these tough-to-beat diseases.

## Example: Beating Biofilms

By working together, bugs [bacteria, yeast, algae, fungi, etc.] can do things they can't achieve alone.... Sometimes this is done in a cooperative, community-spirited way. Other times the newcomers just take advantage, use and trample over and destroy the earlier settlers.

—Dr. James W. Fairley, fellow of the Royal Society of Medicine

Every time you brush and floss your teeth, you are disrupting one of the cooperative wonders of the world: dental plaque. Plaque and other so-called biofilms are colonies of bacteria that contribute personal resources to build up the matrix that binds them together. Biofilms are highly resistant to antibiotics, making them a major public health concern.[9] Fortunately, scientists have recently identified a new and effective strategy by which to "sabotage" biofilms. The idea is to introduce into the film "cheater bacteria" that are genetically designed to be freeloaders who don't contribute to the public good.[10] Once these freeloaders are in place, doctors withhold antibiotics. Absent treatment and free to reproduce, the film grows as all bacteria increase in number. However, the freeloader bacteria grow faster than the rest, since they can devote all of their resources to reproduction.[11] Before long, the freeloaders take over the colony and the film begins to break down due to poor maintenance. At that point, the film becomes vulnerable to antibiotics, which can then eradicate the entire colony.

## The Triumph of Tit-for-Tat

I never expected to find wisdom or hope for the future
of our species in a computer game, but here it is,
in Axelrod's book. Read it.

—*Lewis Thomas on Robert Axelrod,* The Evolution
of Cooperation *(1984)*

In the late 1970s, political scientist Robert Axelrod wanted to know how game theorists would themselves play the repeated Prisoners' Dilemma. So, he announced an unusual "tournament" and invited the world's leading game theorists to participate. Fourteen took the bait, each submitting instructions to program a computer to play the repeated Prisoners' Dilemma on his/her behalf. (Strategies were labeled as "Cooperate" and "Defect," with payoffs in each round as shown in figure 21. Each player's goal was to maximize his/her overall payoff, over many rounds.)[12] As you can probably imagine, most of those who responded to this invitation submitted very complex

**FIGURE 21**    Payoff matrix (each round) for
                       Axelrod's Prisoners' Dilemma

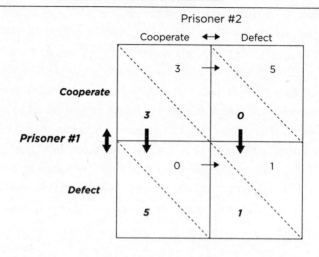

strategies, based on whatever theories they had developed to predict competitors' likely behavior. However, the winner was actually the simplest of all those submitted, a strategy known as Tit-for-Tat.

After Axelrod's initial tournament made its splash and game theorists the world over learned of Tit-for-Tat's success, Axelrod announced a second tournament exactly like the first. This time, sixty-two game theorists threw their hats in the ring. But, guess what? Tit-for-Tat won again, even against this "new and improved" competition!

### TIT-FOR-TAT: QUICK TO TRUST, QUICK TO PUNISH, AND QUICK TO FORGIVE

The Tit-for-Tat (TFT) strategy is defined by two simple rules:

1. Cooperate in round one.
2. Reciprocate after round one; i.e., do whatever the other player did in the previous round.

TFT combines three key features, each of which is essential to its success.

First, TFT is *quick to trust*, as it cooperates in round one. Why is it essential to trust the other player initially? Consider the alternative of a more jaded strategy that (say) waits until the other player proves to be trustworthy before cooperating itself. Relative to TFT, such a strategy has the advantage of avoiding the short-term pain of betrayal in round one if the other player defects. Should the other player cooperate, however, such a strategy has the downside of discouraging the other player from ever trusting *you* again. This could set back any hopes of reaping the long-term rewards of a cooperative relationship. As long as this long-term gain outweighs the short-term pain of being cheated once, it is wiser to give others the benefit of the doubt . . . at least once.

Second, TFT is *quick to punish*, as it responds to any defection by itself defecting in the next round. The threat of punishment is obviously essential, as it incentivizes the other player to cooperate.

Finally, TFT is *quick to forgive*, as it requires the other player to cooperate for just one round before "making amends" and returning to cooperation itself. By contrast, strategies that "hold a grudge" are self-defeating, as they close off the possibility of recovery and reform. Indeed, to incentivize good behavior, it suffices for your punishment to last just long enough to inflict greater pain on the other player than the gain they enjoyed from cheating in the first place.[13] That way, TFT teaches the other player that "cheaters never prosper," but offers forgiveness as soon as that lesson is learned.

## The Evolution of Cooperation

Imagine that Tit-for-Tat has become sufficiently established in a community to be the "strategic norm." Although most people play TFT, there will always be some who go against the grain and adopt strategies that differ from the norm. If those who adopt these "mutant" strategies are more successful than those who obey the norm, the mutant strategy will tend to grow and be imitated more and more, until it becomes the new norm. Should we expect TFT to be a stable norm, resistant to such competition, or will its prevalence in the community likely be more fleeting? The good news is that, in fact, other strategies cannot outperform Tit-for-Tat in a community filled with TFT players.[14] So, TFT is stable as a strategic norm, once it is established in a community.

Stability is nice, but how can a community get to the point at which TFT is prevalent enough to become a strategic norm? Fortunately, there is a natural mechanism by which TFT can successfully invade other, less cooperative strategies. In *The Evolution of Cooperation*, Axelrod explained how with the following colorful illustration, presenting business school education as one possible catalyst for such change:

> A world of meanies can be invaded by a cluster of TFT—and rather easily at that. To illustrate this point, suppose a business school teacher taught a class of students to initiate cooperative behavior in the firms they join, and to reciprocate cooperation from other firms. If the students did act this way, and if they did not disperse too widely (so that a sufficient proportion of their interactions were with other members of the same class), then the students would find that their lessons paid off.

What's notable here is the centrality of *clusters* in the evolution of cooperation. People in a tightly knit cluster have the most to gain from switching to a strategy like Tit-for-Tat, since many of their interactions are with cluster-mates. However, once a cluster converts to playing TFT, that's not the end of the story. As long as players belong to multiple clusters, as is common in social networks, one TFT cluster will automatically *seed* many other clusters to then also switch to TFT, and so on until "everyone who is anyone"[15] has switched to TFT and they all enjoy the fruits of cooperation.

It's easy to see how one reviewer of *The Evolution of Cooperation* could find "wisdom" and "hope for the future of our species" in Tit-for-Tat. After all, the evolutionary drive toward cooperation means nothing less than that the "arc of history" really does bend toward justice, as Reverend King said, as well as toward prosperity and peace. If that's what strategic evolution can do, automatically and without intention, imagine what *people* can do if we put our minds to it.

Consider again Axelrod's example of the business school students whose professor trained them to play Tit-for-Tat. If these students interact enough among themselves, they are better off playing Tit-for-Tat with everyone, compared to always defecting with everyone. But what if they wind up being more widely dispersed and rarely interact? They can still take advantage of Tit-for-Tat, but, to help them do so, the professor's lesson needs to be a bit more nuanced:

1. *Observe* who also knows about Tit-for-Tat.
2. *Teach* others in your strategic network about Tit-for-Tat's merits.
3. *Play* Tit-for-Tat only with those you trust to do the same.

Anyone who sticks to this plan will automatically outperform the defectors of the world, and outperform them more as more others in their strategic network do the same. That's why it's so essential to teach those around you to be quick to trust, quick to punish, and quick to forgive, just as you are. The more you can spread this word, the more complete your success will become. And the more complete your success, the more others will seek you out to learn your secret, allowing you to teach still more and succeed still more.

In this way, just one Game-Changer can be enough to seed and transform an entire organization into a more productive, happier, and altogether better place. Indeed, as Lao Tzu famously wrote over 2,500 years ago, just one could be enough to change the world:

*Realized in one man, fitness has its rise;*
*Realized in a family, fitness multiplies;*

*Realized in a village, fitness gathers weight;*
*Realized in a country, fitness becomes great;*
*Realized in the world, fitness fills the skies.*
*And thus the fitness of one man*
*You find in the family he began,*
*You find in the village that accrued,*
*You find in the country that ensued,*
*You find in the world's whole multitude.*
*How do I know this integrity?*
*Because it could all begin in me.*

> —*Poem 54 of* The Way of Life According to Laotzu *(1944), a poetic rendering of the* Tao Te Ching *by Witter Bynner*

# How to Escape
# the Prisoners' Dilemma

Before proceeding to the applications of Part Two, let's pause for a moment and review the various ways we have seen to escape the Prisoners' Dilemma. The flow diagram of figure 22 summarizes which (if any) escape routes are open to players, depending on the details of the situation.

First, are the players capable of *changing their payoffs*, either directly themselves or indirectly via the intervention of some third party? If so, obviously, they can change the payoffs so that the game is no longer a Prisoners' Dilemma. For example, colleges formed the NCAA to institute new rules to eliminate their incentive for extreme violence on the football field. See chapter 2.

Second, are the players capable of *merging or forming a "cartel"*? If so, again obviously, they can merge so as to look after their collective interest. For example, the Big Four in barbed wire merged to form the American Steel and Wire Company. See chapter 3.

Third, does the game have *dynamic moves*? (Recall that a game has dynamic moves if it occurs in real time and both players can observe and quickly react to changes in each other's moves.) If so, a mutual threat to retaliate is enough to escape the Prisoners' Dilemma. For

FIGURE 22   Summary of Prisoners' Dilemma "escape routes"

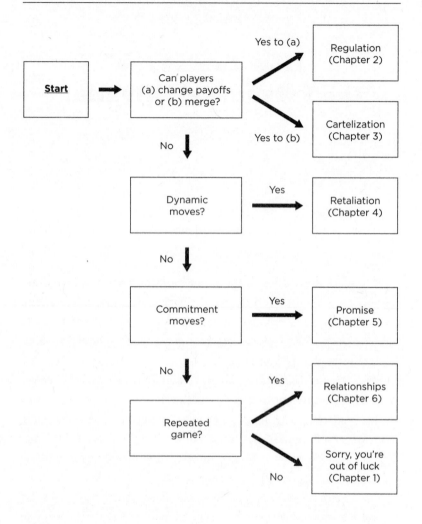

example, in the Dynamic Pricing Game seen earlier (and to be reprised in Game-Changer File 1), the airlines keep posted prices high by each threatening to immediately match any discount. See chapter 4.

Fourth, does the game have *commitment moves*? (Recall that a game has commitment moves if the players move in sequence and the last-mover can commit ahead of time to how it will respond to whatever the first-mover does.) If so, a promise by the last-mover is enough to escape the Prisoners' Dilemma. For example, I get my kids to eat a healthy dinner by promising to serve dessert if they eat their vegetables. See chapter 5.

Fifth and finally, is this a *repeated game*? (Recall that a game is "repeated" if the same players interact repeatedly in the context of a relationship, or, more broadly, if the outcome of the current game can be strategically linked to the outcomes of other games.) For example, the Mafia leveraged their "family ties" to avoid confessing to the police for decades. See chapter 6.

When any of these five conditions holds, there is a way out of the Prisoners' Dilemma. When none of them holds, however, there is no hope of escape. One such hopeless situation is Albert Tucker's classic version of the Prisoners' Dilemma, in which the prisoners are isolated in different cells and asked whether they confess to the crime. This game is clearly not repeated and, since each prisoner is unable to observe the other's decision, the game has simultaneous moves.[1] Consequently, it is not possible to *enable retaliation* (chapter 4), *make a promise* (chapter 5), or *leverage relationships* (chapter 6). If the prisoners could bribe the judge and influence sentences that way, then perhaps it would be possible to *invite regulation* (chapter 2) to change the payoffs. Absent that, however, it's clearly impossible to *form a cartel* (chapter 3) to merge their interests, since each much serve out his own sentence in jail. So, all five escape routes are closed to the prisoners in the classic Prisoners' Dilemma.

Fortunately, these hopeless situations are more the exception than the rule.

Part Two

# THE
# GAME-CHANGER
# FILES

The central claim of this book is that cultivating your game-awareness and mastering the art of game-changing will allow you to identify the essence of any strategic quandary and find ways to change the game to your advantage. Of course, the proof is in the pudding. Here it is, in the Game-Changer Files.

When I began writing this book, I had no idea that in-depth applications would be part of the mix. But as I cast my net in search of interesting examples, everywhere from the business world to the microbial world, I came across several fascinating, important, and apparently unresolved strategic challenges. I couldn't resist diving in and, months later when I finally emerged, the Game-Changer Files were born.

1. *Price Comparison Sites*: How can we help ensure that online competition leads to low prices?
2. *Cod Collapse*: How can we avoid catastrophic regulatory failures in fisheries management?
3. *Real Estate Agency*: How can we reform real estate agency, to better serve buyers and sellers?
4. *Addicts in the Emergency Department*: How can we relieve emergency departments of the burden of drug addicts seeking narcotic pain medication?
5. *eBay Reputation*: How can we enhance buyers' and sellers' trust on eBay?
6. *Antibiotic Resistance*: How can we reverse the frightening global trend toward antibiotic resistance that, if left unchecked, could leave us essentially defenseless against terrible diseases such as tuberculosis?

Every strategic challenge is unique, but a holistic view of the strategic ecosystem is always essential in understanding the crux of the problem. Once such "game-awareness" is achieved, the game-changing principles developed in Part One can be deployed to focus and harness one's creativity and specific knowledge to identify an actual solution. That's what I did in these Game-Changer Files. You can do it too, once you master the lessons of this book.

*Note to readers* For updates on the Game-Changer Files, and more, please visit McAdamsGameChanger.com.

# Price Comparison Sites

Tracking down the best prices for everything you buy can
save some serious cash in tough economic times.

—SmartMoney, *October 10, 2008*

Millions of Americans have discovered the power of smart-phone apps like Amazon Price Check and Internet price comparison websites like PriceGrabber.com. Before purchasing anything from a brick-and-mortar retailer, one can now easily check whether it is available online at a lower price. This phenomenon has even gotten a name, "showrooming," which reflects the fact that brick-and-mortar stores offer an easy way for consumers to view and inspect goods that they intend ultimately to buy online at the lowest possible price.

Showrooming is a mortal danger for some retailers, since online sites like Amazon don't need to maintain a costly retail presence and hence can typically afford to offer lower prices on the most popular products.[1] Yet for all the harm that online price comparison shopping may do to big retailers like Target, or to small mom-and-pop operations (some of whom view Amazon's Price Check app as "kind of evil"),[2] it may seem obvious that shoppers are better off being able to compare prices. However, it's actually not so clear, since the way that firms set prices will itself change as price comparison sites come to permeate the shopping experience.

In some markets, there is compelling evidence that online price comparison shopping has induced firms to set lower prices. For instance, economists Jeffrey Brown and Austan Goolsbee examined the effect of online comparison shopping for life insurance during the 1990s.[3] They found that, as Internet usage spread, prices fell by 8–15 percent for policies that could be shopped online relative to those that could not. (The price of life insurance depends on health characteristics of the insured, e.g., age and smoking history. Only the most common combinations were available on leading online price comparison sites.) Based on this analysis, Brown and Goolsbee concluded that Internet price comparison shopping saved consumers at least $115 million per year.

# The Airline Tariff Publishing Company

In other markets, there is evidence that firms have used price comparison sites as a means to raise prices. The most famous example is the Airline Tariff Publishing Company (ATPCO), jointly owned by several major airlines. ATPCO is the authoritative and exclusive provider of up-to-the-minute fare information, rightly describing itself as "the world leader in the collection and distribution of fare and fare-related data for the airline and travel industry."[4] Consumers routinely access ATPCO fares via a travel agent or online sites such as Expedia or Travelocity. However, consumers aren't the only ones watching. ATPCO also provides fares directly to the airlines, allowing them to observe and quickly respond to one another's moves. Such quick responses could, at least in theory, allow airlines to avoid the low prices normally associated with an intensely competitive industry.

### THE AIRLINE DISCOUNT GAME

To see how this might work, consider a stylized Airline Discount Game played by two airlines that dominate a given route. Each airline must decide whether to offer a discounted fare. Each airline enjoys the greatest possible profit (best outcome) if it is the only one to discount, but suffers the lowest possible profit (worst outcome) if the other airline is the only one to discount. Further, both airlines are better off when neither discounts, compared to when both discount. Note that each airline has a dominant strategy to discount its price, but both airlines are worse off when they both discount. So, the Airline Discount Game is a Prisoners' Dilemma.

If a traveler could solicit secret quotes from the airlines, the logic of dominant strategies would tend to drive both airlines to offer competing discounts. However, the ATPCO system only offers public fares, which all airlines can also see. A price cut by one airline can therefore be immediately matched by others. Forced to choose between all having high prices, or all having low prices, each airline will choose to keep prices high—escaping their dilemma and avoiding discounts.

**FIGURE 23**   Payoff matrix for the Airline Discount Game

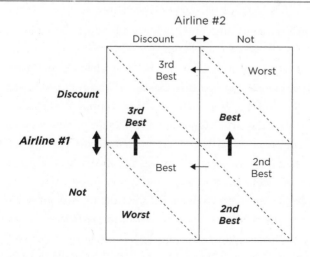

*The ATPCO system only offers public fares, which all airlines
can also see. A price cut by one airline can therefore be
immediately matched by others. The threat of such retaliation
allows airlines to escape the Prisoners' Dilemma of price
discounting.*

This is more than just a theoretical concern. In the early 1990s
(under George H. W. Bush), the Justice Department sued ATPCO and
several major airlines for "price fixing," an antitrust violation under
the Sherman Act. Publicly available materials from the case provide a fascinating window into how airlines used ATPCO in the late
1980s. For example, consider the following back-and-forth between
American and Delta in April 1989 over fares on the Dallas–Chicago
route. (The quote below has been edited and condensed.[5] Note: At
the time, airlines' fares on ATPCO specified a first and last ticket
date, indicating when the fare would be available. Setting the first
ticket date in the future allowed firms to signal fares not yet available for purchase.)

In April 1989, American offered discount fares between its hubs
in Dallas and Chicago on a few select flights each day. Delta
observed American's fares but decided to offer the discount fares
on all of its Dallas–Chicago flights. American then took a number of actions. First, American matched Delta by filing discount
fares on all of its Dallas–Chicago flights, but it added a last ticket
date to those fares of only a few days away, communicating that
it did not want the fares to continue on all flights. American also
refiled the discounts restricted to two flights, with a first ticket
date in the future, thereby telling Delta that American wanted
the availability of the discounts to be limited. At the same time,
American filed fares between Dallas and Atlanta, two of Delta's
hubs, using the same fare levels and last ticket date that it used
on Dallas–Chicago. American thus linked the fares in the two
city pairs, and communicated its offer to withdraw the fares in

Dallas–Atlanta if, and only if, Delta restricted the availability of its fares in Dallas–Chicago.

A Delta pricing employee noted that American's fares in Dallas–Atlanta were an "obvious retaliation" for Delta's fares in Dallas–Chicago. Delta immediately accepted American's offer by withdrawing its discount fares in Dallas–Chicago and filing discount fares that were restricted on two specific flights. American then withdrew the discounts from Dallas–Atlanta. This agreement between American and Delta raised the price of a roundtrip ticket between Dallas and Chicago by as much as $138 for many travelers.

The Justice Department settled the ATPCO case with a consent decree, in which the airlines accepted no guilt but agreed to end certain practices that the government deemed to be anticompetitive. First, the airlines agreed to make most price increases effective immediately, a departure from the previous practice of announcing future price increases and then "negotiating" with the other airlines before finalizing what prices would really be available to consumers. Second, ATPCO itself agreed to change its system to restrict the "footnotes" and other information that airlines could communicate via ATPCO fares. (Previously, airlines had used footnotes to convey how some fares were linked to others. For instance, in the example above, American's retaliatory discount on Dallas–Atlanta contained the exact same footnote information as its fares on Dallas–Chicago, making it all the more obvious to Delta's pricing analysts that the fares were strategically linked.)

Even with this consent decree, of course, airlines can still see and quickly respond to one another's fares, potentially dampening price competition. Overall, then, it's unclear whether the pricing transparency created by ATPCO is good for consumers.

# PriceGrabber.com

What about online price comparison websites like PriceGrabber.com? Unlike ATPCO, these sites typically aren't owned by those who sell the goods whose prices they display, and they actively compete for users by enhancing the quality and clarity of their search results. For example, PriceGrabber calculates the "bottom line price," including taxes and shipping, countering some online retailers' efforts to obfuscate their true prices.[6] Nonetheless, there is little doubt that these sites have the effect of coordinating online retailers' prices. For instance, on Tuesday, May 8, 2012, a PriceGrabber search for "Canon EOS Rebel T3 Black SLR Digital Camera Kit w/ 18-55mm Lens" returned a lowest price of $499 for a new camera. Further, three out of four "featured" retailers—Amazon.com, B&H, and Abe's of Maine—offered exactly that price. (PriceGrabber makes its money by charging retailers each time someone clicks on their link. Featured retailers pay extra to appear more prominently in PriceGrabber's lists.)

Offering exactly the same price is a bit suspicious, but also potentially consistent with fierce competition. To dig deeper, I monitored PriceGrabber listings for the Canon EOS Rebel T3 over several days in May 2012. Prices were mostly unchanging. However, there was *one day* on which *one seller* offered a lower price than Amazon and the rest. On Monday May 7, BestPricePhoto.com ("BPPhoto") offered a price of $409, but only during a "Blow Out Sale!" lasting until midnight.

BPPhoto does not appear to be one of PriceGrabber's most reliable partners.[7] For one thing, PriceGrabber got BPPhoto's price wrong, listing it as $386.43 instead of the true price of $409.[8] It's reasonable to suspect, then, that Amazon's pricing analysts may not even have been aware of BPPhoto's one-day raid into its Canon EOS territory; if they had been, they would have contacted PriceGrabber to revise BPPhoto's displayed price upward to the true price of $409. Even if Amazon were aware, what could it do?

BPPhoto's listing did in fact disappear at midnight, meaning that any retaliation by Amazon would only serve to staunch one day's worth of losses. By contrast, the cost and chaos created by such a retaliatory strike could be considerable. To see why, suppose that Amazon were to match BPPhoto with a price of $409, at the instant at which BPPhoto first offered that deal. The other sophisticated players would immediately see and understand what Amazon was doing and, it's reasonable to suspect, quickly match Amazon's new price as well.

Now put yourself in the shoes of a small camera shop owner who uses sites like PriceGrabber but has little time to think about them. Given Amazon's "price leadership," it would make sense to adopt the following rule of thumb: "I'll check PriceGrabber every once in a while and, when I do, I'll set whatever price Amazon is charging." If a few of these folks visit during Amazon's retaliatory strike at BPPhoto, they will also match at $409, wrongly assuming that this lower price is the "new normal." Even worse, since these sellers won't be checking back for several days and no one wants to move back to $499 until everyone else has done so, the market is now stuck—perhaps for a long time—at the lower price of $409. It's easy to see, then, why Amazon might let BPPhoto's transgression pass.

Now, it's possible that $499 was the "true competitive price" for the Canon EOS camera, and that BPPhoto was just some hapless seller with a camera that it really needed to unload. Or it could be that $409 is still a profitable price at which to sell one of these cameras, but that Amazon has been able to use the dynamic responses afforded by sites like PriceGrabber to maintain prices above their otherwise competitive levels. More research is needed to answer this question, but don't be too surprised if, sometime soon, you hear of a Justice Department investigation into online price comparison sites.

## Changing the Game for Online Retailers

Let's suppose that you're the CEO of PriceGrabber and you are shocked, just shocked!, at the suggestion that online retailers might be using your site to keep prices high. You know it's not true, but you also know that even a completely baseless antitrust investigation could ruin your business. How can you insulate yourself from this risk? Fortunately, once you view this problem through the game-theory lens, it's not hard to identify a few simple tweaks to your business model that will decrease any concern that implicit collusion is rife on your site—or at least make a convincing show that your site isn't designed to facilitate collusion.

Consider the following two-pronged approach. First, rather than allowing retailers to post and change prices anytime they please, only allow retailers to update prices once a day, at midnight.[9] Second, be sure to continue PriceGrabber's current "open access" policy whereby any retailer can sign up to post prices on the site, without any restrictions or punishments related to the prices they offer. Such an approach would eliminate much of the potential concern about implicit collusion, by encouraging new sellers to enter the market should collusion ever take off.

To see why, suppose that I run a scrappy camera outlet that's more than willing to upset Amazon's apple cart, as long as I can earn a few bucks in the process. If Amazon and other big e-sellers were to *immediately* respond to any discount from me, then I would get very few customers as, all else equal, most will prefer to buy from one of the big outlets that have a better reputation. In this way, the current system effectively stacks the deck against new players. On the other hand, if the big guys' ability to respond were delayed, even by a day, I could convert some sales during that time, earn at least a small profit, and (perhaps most important) build up my own reputation for excellent service and eventually join the big-guy ranks.

Of course, having done so, I risk Amazon's wrath. If I were a big outfit, active and perhaps even the leader in several product categories, Amazon could send a clear message by targeting its own discounts in these categories—just as American Airlines targeted Delta's bread-and-butter Dallas–Atlanta route after Delta discounted Dallas–Chicago. Such "multimarket contact" can be an effective way for the big guys to police themselves and avoid the temptation to compete with low prices. However, I'm not a big outfit. I'm a small, scrappy competitor—and a moving target. On sites like PriceGrabber, I have hundreds of products to choose from, when it comes to offering discounted prices. As long as Amazon's response is delayed, I can hit-and-run through different categories, earning a bit of profit each time. And Amazon can't credibly hit me back tomorrow, since I could be anywhere by then.

Big players like Amazon have a strong incentive to eliminate this sort of disruptive competition by small players and new entrants. Perhaps the easiest way to do so would be to convince PriceGrabber to kick such nuisance competitors off the site. And it won't take much to incentivize PriceGrabber to do so. After all, the little guy's business model is based on stealing clicks from the big guys, and PriceGrabber gets paid more when someone clicks on featured sellers like Amazon. That's why "open access" is essential when it comes to fighting implicit collusion on price comparison websites. And that's why the Justice Department could naturally view any move to restrict seller participation on these sites as a fairly clear indication of anticompetitive intent.[10]

# Cod Collapse

For centuries, Newfoundland cod were a wonder to behold. The first Europeans to set eyes upon them, in 1497, told how "the sea there is full of fish that can be taken not only with nets but with fishing-baskets." A hundred years later, English fishing captains reported cod shoals "so thick by the shore that we hardly have been able to row a boat through them."[1] By the early twentieth century, this fishery still produced an abundance, nearly one million tons per year. The only real limit to the catch was people's appetite for cod which, due to Newfoundland's northern remoteness, few ever tasted fresh. Unfortunately for the industry, every known method of preserving the fish altered its taste. That all changed in 1924, when Clarence Birdseye introduced flash freezing to the world.

Birdseye was an American naturalist, working in the Arctic, when he observed how Native Americans there preserved fish. A combination of ice, wind, and subfreezing temperatures instantly and completely froze just-caught fish. Even better, when this fish was cooked and eaten, its taste and texture were the same as when fresh. Birdseye found that flash freezing occurs too quickly for ice crystals to form,

**FIGURE 24**    Annual catch of Atlantic cod and herring, 1950–2010

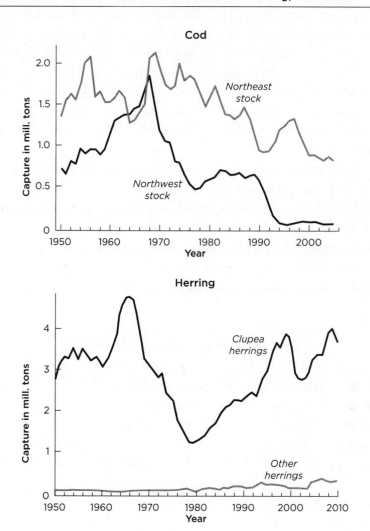

and it's these crystals that destroy the cellular structure of the fish and rob it of its taste. This discovery would transform the world's fishing industry, allowing any fish to be taken anywhere and eaten anywhere else *as if* fresh from the net.

By 1944, Birdseye's company was leasing refrigerated boxcars to transport frozen foods all over the country. The American diet was transformed as millions could now eat fresh fish and, even more important, a variety of fresh vegetables.* With demand stoked by flash freezing, fishermen invested in huge trawlers that finally pulled more cod out of the sea than the fishery could sustain. Not surprisingly, Northwestern Atlantic cod stocks eventually collapsed (in the mid-1970s) with subsequent catches down to about half of previous levels.

Sadly, this sort of collapse was nothing new to the world's fisheries. Fish populations have a critical mass for quick self-replenishment and, whenever overfishing or other factors reduce a population below that level, continued fishing can lead to a sudden and dramatic drop in fish stocks. For example, overfishing in the waters near Iceland and Norway in the 1960s led to the collapse of both the Atlantic herring and the Northeastern Atlantic cod in the 1970s.

## The Fishermen's Dilemma

Fishing communities are devastated when a fish stock collapses. However, fishermen cannot be relied upon to preserve the fish populations on which their livelihoods depend. The reason is that fishermen are trapped in a Prisoners' Dilemma, each with a dominant strategy to catch as many fish as possible but all worse off when the fish stock collapses from overexploitation. That's why fisheries management relies crucially on regulation, not just on market-based incentives.

---

*Birdseye quickly realized that fish were small potatoes compared to other foods, like vegetables, whose markets would also be transformed by flash freezing. In fact, in the United States, the Birds Eye brand is now primarily associated with frozen vegetables.

*Fishermen are trapped in a Prisoners' Dilemma, leading fish
stocks to collapse from overexploitation. That's why fisheries
management relies crucially on regulation, not just on market-
based incentives.*

After the collapses of Atlantic herring and Northeastern Atlantic cod, European regulatory action managed to restrict the catch of
herring and cod enough to stabilize their populations. Herring even
made a comeback, now producing over three million tons per year. By
contrast, regulation of the Northwestern Atlantic cod around Newfoundland has been an absolute failure. After an apparent leveling of
cod populations in the 1980s, the fishery catastrophically collapsed in
the early 1990s, and regulators imposed a complete moratorium on
cod fishing in 1993.

Unfortunately, it appears that even a moratorium will not be
enough to bring back the Newfoundland cod. Normally, one expects
disrupted predator–prey ecosystems to rebalance themselves naturally once the disruption is removed. For example, if farmers hunt
wolves to low levels, the deer population will explode, providing even
more food than usual for wolves to eat. As long as the farmers stop
hunting before the wolves go extinct, the wolf population will tend to
grow back and, in turn, constrain the deer population. But it doesn't
have to work that way. If deer were to eat something that the wolves
needed to survive, it's very possible that the explosion of deer could
spell the end of the wolves, even if further hunting were banned.

That's exactly what appears to have happened to the Newfoundland cod. As a team of marine scientists reported in a 2005 article in
*Science,* "Removal of top predators from ecosystems can result in cas-
cading effects through the trophic levels below, completely restruc-
turing the food web."[2] In this case, the predator–prey relationship
has actually been reversed. Cod's normal prey (such as herring) have
always fed themselves on cod eggs and fry. Historically, this was little

more than a nuisance to the overall cod population, since cod were so numerous. Once the numbers of cod collapsed, however, this relationship reversed itself. Now, herring are the ferocious predator, gobbling up cod before they can grow large enough to return the favor, and the few adult cod who survive are little more than a nuisance to the overall herring population.[3]

Given this potential for complete, irreversible collapse and the Europeans' success at maintaining their cod stocks, one has to wonder: were Canadian and American regulators asleep at the switch or captured by a shortsighted industry? The worst part of this story in some ways is that, no, they were not. As best I can tell, regulators applied the best available science, in good faith, but botched it anyway. The problem is that they had too much faith in their mathematical models and too little awareness of the real games being played.

In 1957, Ray Beverton and Sidney Holt's seminal book *On the Dynamics of Exploited Fish Populations* provided a powerful analytic foundation for fisheries science and management. "Virtual population analysis" (VPA) became—and remains—an essential tool for projecting fish population dynamics and, ultimately, for deciding fishermen's total allowable catch. Unfortunately, as Beverton himself later emphasized in a keynote speech at the 1992 World Fisheries Conference, traditional VPA methodology is fundamentally flawed. Most importantly, VPA assumes that the catch data fishermen provide is correct, ignoring the reality of the game being played between fishermen and the regulator monitoring the catch.

Fishermen are required to report how much fish they catch, how many of these fish are juveniles, and so on. Of course, since these same fishermen are bound by regulations that restrict what they can catch, it's not surprising that they might misreport. For example, since it is illegal to "discard" dead fish before returning to port, no one ever reports doing so. Yet there is indirect evidence that significant discarding occurred.[4] Similarly, since regulators want most to protect younger fish that are the future of the stock, fishermen have an incentive to misreport the ages of their catch, giving the false impression that fewer juveniles are being caught. Consequently, fisheries regulators were blind—and blissfully unaware of their blindness—until it was too late and the cod were gone.

As Beverton emphasized in that 1992 speech, the best way to avoid repeating this tragedy is to change the game in a fundamental way. Don't rely on fishermen to report their catch; monitor it yourself. Sonar-ranging technologies allow monitoring vessels to directly measure the shoals, giving fisheries regulators reliable real-time data to guide their decisions. For decades, fishermen have used sonar to target and, where unregulated, devastate fish populations the world over. Indeed, it's no coincidence that most of the worst fisheries collapses have occurred after World War II, once sonar became widely available.[5] It's satisfyingly ironic that, in the end, sonar may help save the day.

Progress has been made on this front, but other complementary solutions also present themselves once we consider the fisheries problem in a game-theory light. For example, why is it that some fishermen misreport their catch? Certainly not out of any desire to see fish populations collapse. After all, their livelihoods depend on keeping the fish stocks safe and strong. They misreport their catch because they don't want to be caught breaking the rules. But this is an entirely avoidable problem, created by the combination of regulatory monitoring and enforcement.

## Changing the Game of Fisheries Management

Consider the following idea. Create a separate monitoring-only agency to conduct an ongoing "fish census." This agency would track regulatory violations but not share its data with other agencies, much as the US Census Bureau learns of illegal immigration but does not report respondents' immigration status to authorities. The agency would have teeth, however, in the form of surprise inspections that could be targeted based on irregularities in fishermen's reports. For example, fishermen who do not report discarding but tend to catch more than the normal share of valuable fish would be suspect, and targeted more intensively. On the other hand, those who truthfully report illegal activities such as discarding or over-catch would be left alone since, otherwise, the monitoring agency would lose all credibility and hence all ability to conduct its mission to gather accurate information. (The monitoring agency would still report aggregated information about violations, allowing the enforcement agency to identify and act upon areas of concern.)

Fishermen as a whole might still prefer to underreport their catch to this monitoring agency, if they have a short-term mind-set and want the agency to set higher catch limits in the near future. However, when dealing with the monitoring agency, fishermen cannot act in concert but must decide, individually, whether to "confess" what they have fished. Indeed, the fishermen are put into a sort of Prisoners' Dilemma, in which each has a dominant strategy to truthfully report his catch, even though doing so leads to future catch limits that, while perhaps ecologically optimal, may be too low from the fishermen's economic perspective. This is especially satisfying, since a different Prisoners' Dilemma (regarding how much to fish) is what led to overfishing in the first place.

# Real Estate Agency

Homeowners routinely pay as much as 6 percent of their home's sale price to real estate agents, so there's no doubt that such agents provide a valuable service. Without an agent, your home will not be posted to the Multiple Listing Services (MLS) that buyers' agents access when deciding which houses to show their clients. Consequently, "for sale by owner" homes can be essentially invisible to the many buyers who employ an agent to help in their search. In addition to advertising through the MLS, agents provide advice on pricing and help stage the home attractively, among many other services that make the home-selling process as smooth as possible.

Agents also advise their clients how best to negotiate with buyers, including when to accept an offer and when to hold out for more. Two University of Chicago economists, Steven Levitt and Chad Syverson, suspected that sellers' agents might not always give the best advice.[1] To explore this issue, they gathered data on 100,000 homes that sold in the Chicago area from 1992 to 2002 and noticed an interesting fact: many real estate agents buy and sell houses on their own, as a side business. When selling their own properties, agents presumably keep their own counsel and do whatever they think is best. When helping sell someone else's home, do they give the same advice?

Levitt and Syverson's key finding is that real estate agents' own properties sold more slowly (9.5 days or 10 percent longer) and at higher prices ($7,700 or 3.7 percent more) than similar homes that these same agents helped others to sell.[2] One possible explanation is that homeowners are simply more eager to sell their homes than real estate agents are to unload their investment properties. (This could be especially relevant for credit-constrained homeowners who cannot afford to buy a new house until they sell their current home.)[3] If so, we would expect homeowners to be willing to accept lower prices than agents, and to sell their homes more quickly.

Homeowners' desire to sell is undoubtedly part of the story, but other factors also matter. For instance, Levitt and Syverson showed that home prices varied from city block to city block, even after they controlled for home and neighborhood quality, with prices about 2 percent higher on blocks where most homes sold for similar prices, compared to other blocks where home prices were more varied.[4]

Why might the city block on which a home is located affect its sale price? One possible explanation is that homeowners' *information* about what price they should charge depends on whether they live on a similarly priced block. To see how that might work, imagine two city blocks in the same neighborhood that are similar in all respects *except* that homes on one block (Block #1) sell for very similar prices, say from $490,000 to $510,000, while homes on another block (Block #2) sell for more varied prices, say from $400,000 to $600,000. Homeowners on Block #1 have a good idea what price they should charge, since nearby homes all tend to sell for about the same price, but homeowners on Block #2 will need to rely more on their agents when deciding on price. This creates the potential that agents might use their influence to induce clients who live on Block #2 (but not those who live on Block #1) to set substantially below-market prices.

Of course, this story only makes sense if real estate agents actually want their clients to set below-market prices. A typical agent receives a commission of about 1.5 percent of the sale price,[*] meaning that your agent earns more when your home sells at a higher price. Yet representing you takes time, and time is money. For this reason, the best outcome for your agent is for your home to sell *now*, even at a slightly below-market price, since then your agent can shift focus to cultivate other clients and close other deals.

To flesh this idea out, imagine that representing you costs $100 per week and that an offer comes in at $490,000 for your $500,000 home. If you accept, your agent pockets $7,350. But what if you reject and counteroffer at $500,000? If this gamble pays off and the buyer accepts your counteroffer, your agent only gets an extra $150. Given how much your agent stands to lose if the buyer walks away, compared to how little she stands to gain if the buyer accepts your counteroffer, it's easy to understand how she might be tempted to counsel you just to accept the original offer of $490,000, even if she herself would have taken a more aggressive approach when selling a property of her own. In much the same way, it's easy to sympathize with agents who advise homeowners to set their initial asking prices low enough to attract buyers quickly rather than risk leaving homes on the market for a long time,[5] even if they would have been willing to stomach a longer time on the market for their own properties.[6]

## Changing the Game with Sellers' Agents

There are several ways to alter the standard agency contract in order to better align the financial incentives of homeowners and their agents. For example, consider a cost-plus contract, in which the seller

---

*The customary commission, paid by the seller, is 3 percent to the seller's agent and 3 percent to the buyer's agent. However, agents typically pay half of their commission to the agency that employs them.

pays the agent for time and expenses during the home sale process plus a bonus if the house is sold. With a cost-plus contract, your agent faces no risk of loss from taking on your business, and has less financial incentive to rush you to sell.

Further, under a cost-plus contract, the agent has an incentive to provide more of the "ancillary services" associated with selling a house, whenever it's more efficient for the agent to do so. Under the current system, the best agents routinely advise homeowners to hire painters and home contractors, rent furniture, hire landscapers and gardeners, buy fresh-cut flowers, and the rest. The homeowner then typically manages the contractor and pays for the furniture, flowers, and so on. However, it could be much more convenient and cost-effective for *agencies* to manage and pay for such services.[7] Indeed, since agencies represent many clients and can credibly promise repeat business in exchange for great work at a good price, agencies can exploit their scale and relationships with service providers in ways that individual homeowners cannot.[8]

In this way, cost-plus contracting could actually transform the real estate business itself, adding even more high-value services to the long list that agents already provide. Moreover, providing such services would allow the largest agencies to leverage their scale to offer a "product" that smaller agencies could not hope to match. This would allow large agencies to entrench their market power (legitimately, without any concern of violating antitrust law)[9] and deter entry by scrappy smaller players.

The threat of such entry is a serious concern for most large real estate agencies today since, ultimately, the benefits that they enjoy from having a large network are tenuous and mainly based on homeowners' perception that they offer higher-value service. For instance, RE/MAX can claim that "Nobody in the world sells more real estate than RE/MAX,"[10] but how does that help sell *your* house? One reason might be that RE/MAX's success reflects its "respect for the entrepreneurial spirit," which attracts only the most hardworking agents into its fold.

Indeed, RE/MAX burst onto the real estate scene in the 1970s with a completely new "autonomous" business model, in which agents pay all of their marketing expenses, keep all of their commissions, and just pay RE/MAX a "desk fee" for office space and the privilege of being associated with the RE/MAX brand. Joining RE/MAX is therefore truly a sink-or-swim decision that only the most gutsy and hardworking agents are willing to make. However, if you're looking for gutsy entrepreneurs, why not consider a small startup agency that's unencumbered by desk fees? Such a startup could be even more effective at poaching the best and brightest agents from the mainline agencies.[11]

For these reasons and more, big agencies' hold on the real estate market is much less secure than it seems. As I've argued, a shift to cost-plus contracting could change all that, by giving the best agencies an unbeatable strategic advantage in the provision of high-value ancillary services. This makes it all the more puzzling that agencies haven't made more of a strategic shift in this direction. After all, it wouldn't be difficult for the homeowner, alongside the usual exclusive listing arrangement, to agree to pay the agency for any additional services rendered.

Such a shift could also help address an even more fundamental weakness of the real estate agency business. The level of profits that agents enjoy depends critically on lawmakers' failure to address pervasive anticompetitive practices in the industry.* Fortunately for agents, though, the National Association of Realtors (NAR) stands as a rampart against new legislation mandating more effective regulation. Indeed, with nearly a million dues-paying REALTORS® spread throughout every congressional district in the United States, the NAR is a political force to be reckoned with.[12] Even the most obvi-

---

*Dear NAR: Please notice that nowhere do I suggest that real estate agency practices are illegal or even unethical. They're just anticompetitive. I know that sounds pejorative, but really it's not. Indeed, if you've been reading the entire book rather than just this chapter, you'll know that competition (writ large) is squarely to blame for many of the world's ills, including the destruction of one of the richest fishing grounds ever known to man. In those cases, truly, anticompetitive practices would have been a blessing for us all.

ously anticompetitive practices are untouched and, indeed, so universal that most people accept them without a second thought. (More on this later.)

## The Game with Buyers' Agents

You've just accepted a job offer in a new city, and fly with your family to get the lay of the land. Your boss introduces you to a real estate agent, who will spend the day showing you neighborhoods and houses. But what houses will she show? To answer that question, you need to put yourself in the agent's shoes and consider what she knows about the various houses on the market. First, any good agent knows that houses have "personalities," and will seek to match you accordingly. Indeed, many factors go into making a home a good fit for your family, everything from the price and number of bedrooms to the nearby schools and restaurants. But there's another crucial factor that influences whether a house is also a good match *for your agent*: the commission percentage being offered by the seller.

Let's imagine that your new job is a lucrative one, and you're looking at million-dollar mini-mansions. If the agent shows you a home with a 3 percent commission and you buy it, she and her agency will pocket $30,000. However, if the agent shows you a home with a 2 percent commission and you buy it, their haul falls by $10,000. Clearly, then, you are more likely to see homes with a 3 percent commission! Recognizing this, agents on the seller's side routinely counsel homeowners to be sure and offer the buyer's agent the full customary 3 percent commission, if they want to have a shot at attracting buyers represented by an agent.

What's really going on here? Homeowners are locked in a Prisoners' Dilemma—the Agent Commission Game of figure 25—in which each homeowner has a dominant strategy to offer the full commission of 3 percent to the buyer's agent rather than a lower rate of (say) 2 percent.

**FIGURE 25  Payoff matrix for the Agent Commission Game**

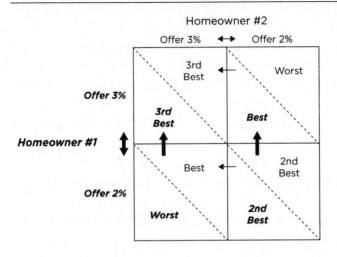

This Prisoners' Dilemma hinges on an exceedingly strange practice that nowadays everyone takes for granted: the homeowner and his agent decide how much the buyer's agent will be paid for services rendered to the buyer. Outside of arranged marriage, it's hard to think of another situation in which two parties determine the terms of trade between two entirely different parties. And arranged marriage at least has something going for it, since parents have a natural incentive to look out for the well-being of their children. By contrast, in real estate sales, no one is looking out for the buyer.

The commission paid to the buyer's agent doesn't come out of the buyer's pocket directly. However, the buyer pays this fee indirectly, since the seller internalizes the commission that will have to be paid. For instance, if the buyer's agent will be receiving $10,000 more in commissions, the homeowner will demand $10,000 more in price. In this way, the commission paid to the buyer's agent is ultimately "passed through" to the buyer. Going back to our original example, that day spent driving around to look at million-dollar houses could wind up costing $30,000—quite an expensive ride.

# Changing the Game with Buyers' Agents

Why not let the buyer and his agent negotiate over how much that ride should cost? Buyers still need real estate agents to help them find a good house and navigate the byzantine process of purchasing a home. Undoubtedly, they'll still pay handsomely for a good agent's help. However, they probably won't pay as handsomely as before, because agents will have more incentive to compete for clients.

Forcing buyers' agents to negotiate directly with their clients about how much they will be paid fundamentally changes the game that sets these agents' fees. In particular, the Agent Commission Game of figure 25, played by homeowners, is transformed into the Agent Competition Game of figure 26, played by buyers' agents. In this game, agents must compete to represent a client.[13] Even if the buyer is happy to pay the full commission of 3 percent, he would obviously prefer to pay the lower rate of 2 percent. Eager to acquire the buyer's business, agents have an incentive to compete aggressively, distinguishing themselves by offer-

**FIGURE 26**  Payoff matrix for the Agent Competition Game

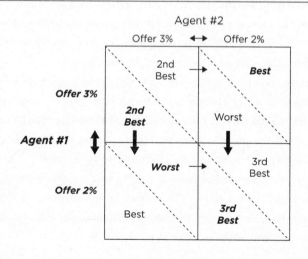

ing lower fees (and better service). Indeed, as figure 26 illustrates, this game is itself a Prisoners' Dilemma, in which each agent has a dominant strategy to discount but all agents are worse off when all discount.

Buyers are better off once the game is changed in this way, for several reasons. First, as already discussed, agents will be forced to compete. This induces agents to charge lower prices, as well as to provide higher-quality services to buyers. Perhaps even more important, however, is the impact that such a change would have on the real estate business itself. Buyers aren't willing to pay thousands of dollars just to be driven around and shown where to sign. Consequently, agents who don't provide outstanding service to buyers won't stand a chance. Indeed, it's easy to imagine the rise of a new breed of real estate agent who specializes entirely in serving buyers—and earns good money because they do it so well.

## Allow Commissions to Reflect the True Value of Services

As the housing market emerges from its collapse and long doldrums, we have a historic opportunity to focus on real estate and "fix the game" once and for all. Inefficiencies in real estate agency are just one piece of the problem, but a significant one. Agent commissions represent a "friction" that likely deters at least some homeowners and potential homeowners from entering the market, ultimately contributing to a lack of liquidity and an inefficient allocation of our housing stock. We need a system in which agent commissions transparently reflect the true value being offered to buyers and sellers.

A natural first step would be for lawmakers to bar homeowners from paying buyers' agents. Such payments create a clear conflict of interest and choke the American real estate market of its natural dynamism. Some REALTORS® may go apoplectic at this idea, but similar arrangements are already banned in many other industries, for good reason. For instance, how would you feel if your heart surgeon received most of his income from pacemaker makers, who paid him every time he installed one of their devices? No doubt your surgeon would keep your well-being foremost in mind and pick the pacemaker that he thought was best for you. However, I guarantee it: your surgeon would take the money too, as long as it was legal to do so.

*Lawmakers ought to bar homeowners*
*from paying buyers' agents.*

Even if no surgeon were ever tempted to shortchange a patient on quality, such side payments would still create a significant barrier to entry for new lifesaving products and technologies, allowing the powers that be in the medical device industry to entrench their market dominance. There's a good reason why such "pay-to-play" payments are banned by law.[14] Even though side payments would sweeten the deal for surgeons and encourage more people to pursue that career path, arguably a good thing, they would also undermine competition and the inventive dynamism that goes along with it, a far worse thing.

In much the same way, there's no sense in arguing that many real estate agents *need* their 3 percent commissions and can't get by on less. There's no denying that regulatory changes leading to lower commissions for lower-performing agents would create real pain, as many thousands of real estate agents would leave the business to pursue other careers. However, the cold hard fact is that it's relatively easy to become a real estate agent, and such "free entry" will always keep some agents' earnings low, whatever the regulatory regime. The main difference, when all is said and done, is how many people choose to be real estate agents. Fewer agents means fewer dues and less clout for the NAR, but that's not necessarily a bad thing for homeowners and home buyers, nor for the agents who remain. Indeed, as I have argued, unleashing true competition could transform the real estate business in such a way that the best agents and the best agencies could be very profitable indeed.

# Addicts in the Emergency Department

> The overuse of narcotics is a huge problem, and when a patient presents, especially for dental pain, it's difficult to make an objective assessment.... We err on the side of treating pain, and it is a huge potential for abuse.
>
> —*Gail D'Onofrio, Professor and Chair, Department of Emergency Medicine, Yale School of Medicine, 2012*[1]

> Ask emergency physicians where their greatest job stress comes from, and it is unlikely they will name the trauma, violence, or death that are a part of this vocation. Instead, frustration and fatigue come from the barrage of daily visits from patients seeking relief from pain.
>
> —*Charles Fooe, MD, et al., in* Emergency Medicine News, *2011*[2]

> I know I'm addicted to [opioids], and it's the doctors' fault because they prescribed them. But I'll sue them if they leave me in pain.
>
> —*A patient as quoted by Anna Lembke, Assistant Professor of Psychiatry, Stanford University, 2012*[3]

When accident or sudden illness strikes, quick access to the emergency department (ED) of your local hospital is often your only option—and sometimes your only hope. Unfortunately, in America today, there is simply not enough space to serve everyone in urgent need. Indeed, about once a minute, an ambulance is diverted to a more distant hospital because the closest ED cannot handle any more patients. The Institute of Medicine has called this overcrowding problem a "national epidemic," one that adversely impacts the quality of care when we need it most.[4]

Many factors contribute to the overcrowding of America's emergency departments. Some problems are structural, like the inadequate number of non-ED hospital beds into which to discharge patients, but others are strategic and hence potentially amenable to solutions that "change the game." For instance, patients without health insurance may *choose* to forgo prevention or delay early treatment of a disease because of cost considerations, only to wind up later in the emergency department, where much of the cost of their care is ultimately shifted to the general public. There are many ways to change that game, e.g., by lowering the cost of health insurance coverage, directly by subsidy or indirectly by increasing competition among health insurers. However, the health care system is so complex, and any reform so rife with unintended consequences, that it can be difficult to identify the best path forward... and well nigh impossible for politicians to agree on a course of action.[5]

Fortunately, it's possible at least to mitigate emergency department overcrowding by focusing on another contributing factor that is much simpler to grasp: addicts looking for drugs. Emergency physicians are continually beset by patients who claim to be in severe pain. The pain of some of these patients is real—one common culprit is an infected tooth nerve—but others are addicted to pain medication (especially opioids) and are simply seeking a fix. Unfortunately, the number of such drug-seekers appears to be on the rise. Indeed, according to the Centers for Disease Control and Prevention, the number of ED visits linked to the nonmedical use of prescription pain relievers more than doubled from 2004 to 2008, from 144,644 to 305,885 visits a year.[6]

# The Emergency Physicians' Dilemma

This backdrop of widespread drug-seeking behavior puts emergency physicians in a difficult position when a patient presents who claims to be in severe pain and (say) insists that he is allergic to all pain relievers *except* addictive ones like Percoset or Vicodin.[7] Pain and allergies are not directly observable, so there is no way to know for sure if a patient like this is lying. Plus, even if doctors could reliably spot deceit, the time it takes to separate the drug-seekers from the pain sufferers could be used to help those in more urgent need of medical attention.

Denying pain medications to obvious drug-seekers undoubtedly gives some emergency physicians peace of mind that they are not feeding dangerous addiction to narcotics. Yet there is reason to suspect that such peace of mind may not be a strong enough motivation for at least some doctors. For one thing, a drug-seeking patient that one doctor turns away can simply try his luck again, at a different time or at a different hospital. And if the next doctor will give him his meds anyway, what's the use of bothering to deny him yourself? Dr. Thomas Benzoni, a longtime emergency physician in Sioux City, Iowa, recently explained this logic to the *New York Times*:

> I admit that some people get drugs out of me who shouldn't get them. [But what am I supposed to do?] Do I deny them drugs so that one person doesn't get a little more Vicodin? It's [like] emptying the ocean with a teacup to try to address our societal drug problem.[8]

Altogether, emergency physicians appear stuck in a Prisoners' Dilemma, each with a dominant strategy to prescribe pain medication to everyone who requests it, but all worse off as such easy access to narcotics sustains addiction and encourages more and more drug-seekers to flock to *every* ED.

An emergency department could shape this game by adopting policies that make itself less attractive to addicts. If so, drug-seekers would eventually learn to avoid that particular ED, thereby shortening wait times for patients with more urgent needs. In reality, however, most EDs have adopted policies that further incentivize doctors just to hand out drugs. For instance, emergency physicians nowadays are routinely rated on "patient satisfaction," with their pay and even job security tied to satisfaction scores. Not surprisingly, some emergency physicians don't bother denying drugs even to obvious drug-seekers. (More on this later.)

## Changing the Game of Emergency Pain Relief

Whatever the challenges, there's one thing working in our favor as Game-Changers looking to transform emergency pain relief. Everyone agrees that drug-seeking in the emergency department is a serious problem that needs to be addressed. The trouble is that the medical community hasn't figured out how to work together to find a solution. That's where game theory can help, by providing a systematic approach to identify the key strategic interactions, and the key features of the environment, that are blocking doctors and hospitals from achieving what all agree ought to be our common aim.

The problem of drug-seeking in the emergency department can, in particular, be traced to two key strategic factors:

1. *Asymmetric information*: Emergency patients know whether they are in genuine pain or just faking to score drugs, but emergency physicians do not know for sure which patients are really in pain.
2. *Perverse incentives*: Because of pressure from hospital administrators to "satisfy" patients, emergency physicians have an incentive to write opioid prescriptions even for people whom they suspect of being addicted.

Consider first the problem of asymmetric information. Emergency physicians' inability to directly observe whether a patient is truly in pain puts them at a disadvantage, as they cannot detect and deny fakers without also leaving some genuine patients to suffer in pain. Doctors naturally respond to this uncertainty by erring on the side of caution, prescribing treatment whenever there is any doubt, despite the harm that comes from giving addicts easy access to narcotics.

Similar quandaries arise in a wide variety of strategic interactions in which some player's "type" is hidden from others—everything from tax collection (the IRS cannot identify tax cheats without also auditing some honest taxpayers) to the search for a romantic partner (relationship-minded women cannot avoid "players" without also distancing themselves from some desirable men). Unlike in these other settings, however, there is no *inherent* reason why emergency physicians cannot identify patients who are addicted to pain medications. For instance, suppose that doctors could easily track each patient's prescription history. Those with a steady stream of recent prescriptions to opioids might not be addicts—they might need opioids to treat chronic pain—but such a person who shows up in the emergency department claiming to need a few more pills to "tide them over" almost certainly is.

Prescription information is dispersed throughout the health care system, and aggregating it all into a single database is a Herculean task. Fortunately, efforts are already under way in many states to pool such information. Indeed, as reported by the Prescription Drug Monitoring Program (PDMP) Center of Excellence at Brandeis University, forty-four states either had an operational PDMP or had adopted legislation authorizing a PDMP in April 2011.[9] The rise of Prescription Drug Monitoring Programs now allows emergency physicians to weed out many addicts whose prescription histories appear in the relevant PDMP. Furthermore, since ED prescriptions are themselves entered into these databases, this means that addicts (or those scoring drugs to sell to addicts) can no longer *habitually* frequent emergency departments to procure drugs without labeling themselves as addicts.

For this to work, though, emergency physicians must make an effort to check their local PDMP database every time an emergency patient arrives claiming pain. Unfortunately, even that may be too much. Despite the rising availability of comprehensive PDMP databases, few emergency physicians check to see whether patients have recently been given painkillers before writing a new prescription themselves.[10] According to John Eadie, director of the PDMP Center of Excellence, "Many emergency physicians don't realize the importance of a quick check of the database to see how many painkiller prescriptions a patient has filled lately." If Mr. Eadie is correct, a bit of education and an easy-to-use smartphone app could be enough to push emergency physicians over the hump toward widespread PDMP adoption.

Unfortunately, I fear the real problem may be deeper (and more disturbing) than simple lack of awareness. Emergency physicians are savvy individuals who are already quite aware of and deeply frustrated by the problem of addicts in the emergency department. If PDMPs really solved the drug-seeking problem that emergency physicians face on a daily basis, PDMPs would be widely adopted in a heartbeat. However, PDMPs don't *on their own* solve the strategic problem that doctors face when a patient presents with claims of pain.

To see why, just put yourself in their shoes for a moment. Emergency physicians who refuse to prescribe narcotic pain medication to suspected addicts will inevitably upset these "patients" and earn lower satisfaction scores. "You can be faulted for not treating a patient's pain—it's considered the 'fifth vital sign,'" explains Dr. Abhi Mehrotra, assistant director of the ED at the University of North Carolina Hospitals. Or as Dr. Thomas Benzoni put it, even more bluntly: "If you're going to criticize me for not giving out narcotics, and you never praise me for correctly identifying a drug-seeker, then I'm going to give out narcotics." Of course, once you're resigned to just giving out narcotics, what's the point of checking the PDMP to confirm your suspicion that a patient is an addict? Doing so will just make you feel worse about "treating" their pain.

## "Satisfaction" Focus Creates Perverse Incentives

Why don't hospital administrators give their doctors more discretion to refuse to write pain medication prescriptions? First, no one wants to make a mistake and fail to provide relief to someone truly in pain—especially in these litigious times, when "pain and suffering" is cause for monetary damages.[11] Second, hospitals don't like having dissatisfied patients.

Those "I strive for five [out of five]" signs posted all over hospitals these days aren't just window dressing. Hospitals are desperate to distinguish themselves from competing facilities—to attract lucrative "destination shoppers" who can choose among many hospitals for their care—and customer satisfaction rankings provide one of the easiest and most salient ways to do so. For instance, Jack Horner, president and CEO of Major Hospital in Shelbyville, Indiana, crowed in 2009 about his hospital's high scores on a survey run by the healthcare consulting firm Press Ganey:*

We were very pleased to see that Major Hospital ranked No. 1 in comparison to selected similar-sized hospitals in the region and in the top 2% nationally. Press Ganey is a highly respected independent healthcare consulting firm, and their report eloquently confirms the success of the continuous improvement commitment by Major Hospital's physicians, nurses and professional staff. The goal of Major Hospital is nothing less than to be the best in the region for personalized healthcare delivery, and the

---

*Press Ganey occupies a unique, powerful, and highly profitable position within the hospital management industry, and it uses that position to drive mostly positive change in how American hospitals provide care. In 2008, the private equity firm Vestar Capital Partners acquired majority ownership of the firm in a management takeover. Since then, Press Ganey has expanded its formidable reach beyond hospitals to medical practices, outpatient services, and home care.

Press Ganey results demonstrate that we are advancing toward
that goal.[12]

It's fair to say that hospital administrators are fairly obsessed with
rising in Press Ganey's hospital ranking. And with good reason: hospital executives' bonuses are routinely tied to Press Ganey scores
while, not coincidentally, many emergency physicians' pay depends
on their Press Ganey evaluations. Indeed, in one survey of emergency
physicians conducted by the WhiteCoat's Call Room blog of the *Emergency Physicians Monthly* in December 2009, "Nearly one in eight
respondents had their employment threatened due to low patient satisfaction scores."[13]

Doctors naturally respond to these incentives by working harder
to satisfy patients. At first glance, that might seem like a good thing.
After all, higher patient satisfaction does appear to be correlated with
a better quality of care.[14] However, just because higher patient satisfaction is *statistically* linked to better quality of care doesn't mean
that initiatives that increase patient satisfaction will also necessarily increase the quality of care. To see why, allow me to digress for a
moment with an analogy from the world of baseball.

In March 2013, Washington Nationals pitcher Stephen Strasburg expressed his admiration for the stamina of other outstanding
pitchers in the league, saying, "You look at some of the top pitchers
in the game, they [throw] at least 110 [pitches] every time out. I'm
going to be prepared [to throw that many pitches . . . I've] just got to
keep working, keep grinding."[15] It's true that pitchers who give up
the fewest runs per inning also tend, on average, to stay in the game
longer and throw more pitches. However, that doesn't mean that the
best way to improve as a pitcher is to throw more pitches. Indeed, it's
natural to suspect that pitchers who push themselves to throw more
pitches than they normally would are apt to get tired, or injured, and
perform worse than if they had left well enough alone.[16] In much the
same way, doctors who go out of their way to "satisfy" patients may

actually provide a lower quality of care than if they had simply served those patients to the best of their abilities.

Focusing on patient satisfaction may sometimes even give doctors an incentive to provide inappropriate care.[17] In the *Emergency Physicians Monthly* survey mentioned earlier, for instance, "More than 40% of [emergency physicians who responded] had altered treatment due to the potential for a negative patient satisfaction survey. Of those that altered treatment, 67% gave treatment that was probably not medically necessary more than half of the time, [sometimes resulting in complications including] kidney damage from IV dye, allergic reactions to medications, hospital admits for 'oversedation' with pain medications, and Clostridium difficile diarrhea."

# The Wrong Patients Are Being Satisfied

The best way to satisfy patients in the emergency department, it would seem, is to save their lives. Yet when an emergency physician saves a patient's life, s/he typically gets no credit for doing so. The reason is that, after being saved in the ED, the sickest and most seriously injured patients are typically transferred to the hospital's inpatient facility which then gets full credit for their satisfaction. As emergency physicians William Sullivan and Joe DeLucia explained in the September 2010 issue of *Emergency Physicians Monthly*:

> Patients admitted [from the ED] to the hospital and patients transferred to other hospitals do not receive Press Ganey emergency department satisfaction surveys. Some questions about the emergency department may be included on inpatient surveys, but the answers to those questions count toward the inpatient satisfaction scores, not the emergency department satisfaction scores. [This can create a dilemma for emergency department staff.] Should emergency physicians and nurses provide appro-

priate yet time-consuming medical care to high acuity patients, or should they provide a minimal amount of medical care to the sickest patients so that they can focus more attention on patients who will be completing satisfaction surveys? Sometimes, especially in single-coverage emergency departments where staffing has been cut due to budget constraints, "doing both" may not be an option.[18]

Since only the least seriously ill ED patients ever receive an ED survey, Press Ganey's ED scores mainly reflect the experience of non-acute patients, especially how long such patients have to wait until all seriously ill patients have been treated. Hospitals can take steps to limit the wait in their emergency departments and, indeed, many hospitals pay Press Ganey handsomely to advise them how to do so . . . and see their Press Ganey rankings rise as a result. But what does the length of the wait in the ED really tell us about overall hospital quality?

ED wait times are influenced by many confounding factors that have little if anything to do with the overall quality of care that a hospital provides. For instance, a hospital situated in an area with more crime, or more uninsured patients, is likely to have longer waits no matter what the hospital does. Recognizing this, Press Ganey benchmarks hospitals against "similar" facilities, i.e., other teaching hospitals, other urban hospitals, and so on. But hospital environments differ in so many ways, not controlled for by Press Ganey, that even these benchmarked comparisons may mean very little.

Dr. Gail D'Onofrio provides an especially telling example of the limitations of Press Ganey's survey results. Professor and chair of the Yale School of Medicine's Emergency Medicine Department, Dr. D'Onofrio runs two fully staffed emergency departments, each of which receives several tens of thousands of patients per year. The first, located in New Haven's gritty downtown area, routinely ranks in the bottom 20 percent in customer satisfaction among Press Ganey respondents. The second, in the shoreline area of nearby Guilford, routinely ranks in the top 1 percent. Indeed, the difference is so stark that one patient recently complained that the doctors in New Haven should take a lesson from those in Guilford. The only problem? The same emergency physicians serve in both locations! What's different, then? Several things but, as Dr. D'Onofrio told me, a primary factor is that Guilford serves fewer uninsured patients. Plus, there are fewer drug-seekers.

# How Hospitals Can Change the Game

If there's one thing that hospital boards care about more than rising in the Press Ganey rankings, it's avoiding the threat of lawsuits. Unfortunately, in addition to undermining the quality of care in the emergency department, hospitals' widespread practice of incentivizing doctors based on Press Ganey survey results potentially opens hospitals up to an entirely new sort of legal liability. As Dr. William Sullivan and Dr. Joe DeLucia explained, again from the *Emergency Physicians Monthly*:

> If adverse patient outcomes due to unnecessary treatment can be tied to pressures that hospitals place on the medical staff to improve patient satisfaction scores, civil liability to the hospital could result. Knowledgeable lawyers could allege that hospitals or physicians cut corners with critically ill patients in order to focus attention on patients who will be receiving satisfaction surveys. [Moreover,] billing Medicare for treatments or hospitalizations that are provided solely from pressure to improve patient satisfaction scores will likely [result in] increased attention from Medicare RAC auditors [who might seek to claw back past Medicare payments]. A pattern of such overutilization, if substantiated, may be sufficient to warrant sanctions against a hospital.[19]

Hospitals can easily mitigate this legal liability risk by changing how they use Press Ganey survey results. For instance, suppose that hospitals were to remove all financial incentives that encourage ED administrators and emergency physicians to focus on improving Press Ganey satisfaction scores.[20] This step alone would likely immunize a hospital from any legal liability, as it would eliminate the hospital's role in creating emergency physicians' perverse incentives to provide inappropriate care.[21]

The good news for hospitals is that new and better options are coming to help them measure patient satisfaction and learn from patient feedback. In particular, a promising startup named Bivarus appears poised to soon burst onto the scene and challenge Press Ganey's long-held dominance of the patient satisfaction market. Founded by a pair of Duke and UNC doctors in 2010, Bivarus's mission is to provide a more scientifically grounded, clinically based approach to measuring patient satisfaction:

> [Bivarus is] a response to frustration among healthcare professionals with current patient satisfaction tools and their inherent lack of scientific precision and actionable findings. We have created a truly revolutionary, Bayesian-based platform[*] to generate precise findings around the patient experience and, importantly, to provide sound data around key aspects of care to drive on-the-ground intervention for positive change. We believe Bivarus will be the catalyst to propel healthcare into a 21st-century mindset—one where patients are central to the dialogue and their scientifically-validated feedback guides us away from the limitations of mere benchmarking and scoring to sound decision-making through truly listening and honoring the voices of those whom healthcare serves.[22]

Bivarus's pilot program, launched in the UNC emergency department in 2012, has generated a game-changing level of patient engage-

---

*In other words, Bivarus uses a statistical model to decide what questions to ask each patient.

ment, with unheard-of response rates of 30–50 percent across several clinical areas.[23] (Typical response rates for traditional patient satisfaction surveys are less than 5 percent.) With their voices finally being truly heard, patients may indeed become "central to the dialogue," allowing hospitals and providers to harness meaningful patient feedback to improve the ways in which health care is delivered and received.[24]

# Unleashing a "Race to Toughness"

Removing the pressure on emergency physicians to earn positive Press Ganey scores will immediately improve emergency department care in many ways, as emergency physicians are finally freed to treat their patients with candor, dignity, and the best possible care. Unfortunately, the problem of drug-seeking in the emergency department is more deeply rooted, and less likely to improve so quickly. The most egregious drug-seekers may be weeded out, but the subtler ones will always tell a story plausible enough to convince the physician that their pain may be real. For such patients—let's call them "probable drug-seekers"—doctors and hospitals will still have an incentive to err on the side of prescribing narcotics.

That said, a game-theory analysis of the situation provides reason to expect that, over time, emergency departments will adapt to become tougher on probable drug-seekers. To see why, note that freeing emergency physicians and ED administrators to decide how tough they want to be will naturally lead to *variation* in how different EDs treat probable drug-seekers. Some EDs will be more lenient, continuing to give narcotics to essentially anyone who asks, while others will take a harder line, refusing relief to anyone whose story seems suspicious.

Addicts will naturally respond to this variation, eventually focusing their efforts on whichever EDs represent the easiest targets. In this way, the toughest EDs will be rewarded with fewer visits from drug-seeking patients. Wait times at these tougher EDs will improve, not to mention doctor morale, all of which will ultimately translate into greater patient satisfaction as well. In this way, hospital administrators who get tough on drug-seekers will enjoy an advantage relative to their more lenient competition.

Of course, as one hospital gets tough and pushes drug-seekers to frequent its competitors' emergency departments, those other hospitals will face an even greater crush of addicts knocking down their doors. These other hospitals will then have a stronger incentive than before to get tough themselves and, indeed, to get even tougher than the rest. In the end, this dynamic creates a "race to toughness" that will tend to drive all hospitals to be tougher on drug-seekers than they would have chosen to be on their own.

The end result is that drug-seekers may find emergency departments much less hospitable places to procure narcotics. Indeed, with easy access to drug prescription databases *and* vigilant emergency physicians in place, addicts and dealers may find EDs so unattractive that they give up even trying to get drugs there. If so, drug-seeking in the emergency department could finally become a thing of the past, leaving our emergency physicians with time and energy to focus on their real job—saving lives!

## Postscript: A Call for Leadership

Each year, emergency departments in the United States see over a million patient visits linked to the misuse or abuse of prescription drugs.[25] Best of all would be for hospitals to take advantage of these points of contact to *help* addicts escape their dependencies. Unfortunately, as explained above, the natural strategic evolution of the emergency department ecosystem—once hospitals recognize the folly of focusing on "satisfying" their emergency patients and free emergency physicians to provide the best possible care—is toward toughness, not kindness.

With the right leadership, however, hospitals can rise above and change this game, not just to deny narcotics but to empower those with a drug dependency with the information and resources they need to control their addictions. For instance, doctors at the Yale emergency department have spearheaded a model "Screening, Brief Intervention, and Referral to Treatment (SBIRT)" program to provide "behavioral treatments for patients presenting to the Emergency Department who abuse drugs and engage in hazardous drinking. This area is of particular significance . . . as substance abuse problems are a leading cause of preventable illness and injury."[26] Widespread adoption of programs like SBIRT could be a significant turning point in our struggle with drug and alcohol addiction, if only we can find the personal and national will to make it so.

# eBay Reputation

By creating an open market that encourages honest
dealings, I hope to make it easier to conduct business
with strangers over the net. We have an open forum. Use
it. Make your complaints in the open. Better yet, give
your praise in the open. Let everyone know what a joy it
was to deal with someone. Above all, conduct yourself in
a professional manner.

*—Pierre Omidyar, eBay founder, 1996*

During the summer of 2012, as part of its back-to-school promotion, Apple handed out $100 iTunes gift cards to any ris-
ing college student who purchased a Mac computer. Even if
you aren't interested in buying stuff on iTunes, such a gift card is still
valuable. Indeed, a number of these cards sold on eBay, typically for at
least $80. That said, if you're a buyer, beware! The eBay gift card mar-
ket appears rife with unscrupulous sellers.

To get a sense of this market, I examined all $100 iTunes gift cards
that actually sold on eBay between midnight and 2 pm on October 2,
2012. Figure 27 summarizes what I found. eBay offers two ways to sell
an item: by auction or with a "Buy It Now" price that buyers can simply accept. Of thirty-one sales, twenty-two were at a Buy It Now price
*exceeding* the gift card's $100 face value. Why would anyone pay more
than face value? One plausible explanation is that the "sellers" of these
gift cards were conducting phony transactions to boost their standing
on eBay, "buying" their own cards and then giving glowing "feedback."[1]

Five other sales were conducted by auction. One of these was
also clearly phony, attracting just one bid of $99.99 which, once one
includes $3.99 for shipping, comes out at $103.98. The four other

FIGURE 27   Total price (including shipping) on all eBay sales
            of $100 iTunes gift cards, midnight to 2 pm, on
            October 2, 2012

|            | Actual Sale | Unclear | Phony Sale |
|------------|-------------|---------|------------|
| Buy It Now | *3 sales* *(84.95–$90.00)* | *1 sale* *($97.95)* | *22 sales* *($102.99–$109.70)* |
| Auction    | *N/A* | *4 sales* *($91.00–$98.60)* | *1 sale* *($103.98)* |

auctions finished at prices less than $100, but that's no guarantee
that these sales were real. In fact, a clever crook could conceal the
phoniness of a sale by allowing genuine competition but entering a
"proxy bid" himself that exceeds $100. As long as you face any com-
petition for an item, eBay will automatically bid on your behalf up
to your proxy bid. By setting a proxy above $100, a phony seller can
essentially guarantee that no real buyer will ever win the auction.
Moreover, the final price in such an auction won't ever exceed $100,
since all real buyers will drop out at lower prices. This would largely
immunize a crook from suspicion, should future buyers ever exam-
ine his sales history.

That said, there were at least a few actual sales. Four transactions
took place at Buy It Now prices under $100, of which three were sold
at a discount of 10 percent or more off face value.[2] Of course, some of
these sales could still have been fraudulent, if the gift cards in question were used, stolen, or irredeemable for some reason.[3] Indeed,
according to the National Retail Federation, "Gift cards on online

auction sites are more likely to be counterfeit or obtained through fraudulent means.... Shoppers should only buy gift cards from reputable retailers, not online auction sites."[4]

The fact that unscrupulous sellers bother to conduct phony transactions to boost their eBay reputation is clear evidence that reputation matters. The idea most buyers have is that, since it takes time and effort to accumulate positive feedback, a seller with good feedback is more likely to be a legitimate seller. However, as an eBay guide from 2007 titled "Can You Still Trust eBay Feedback?" explains, even scammers can cultivate sterling eBay reputations, the better with which to con unsuspecting buyers:

> The eBay feedback system has long been the backbone of eBay .com. In the past, eBay feedback made it easy to avoid online scammers, if buyers dealt only with sellers with lots of positive feedback. Today, unfortunately, it is just not that simple. Online scammers have found various ways to cheat the eBay feedback system to obtain false positive feedback. The latest craze is the purchase of online recipes, e-books, wholesale lists, free information, and information booklets. Simply put, any item that can and does sell on eBay for under $1 is fair game for scammers looking to purchase positive feedback. It's easy for a thief to buy 10 recipes or e-books (gets a yellow feedback star) for under $1. That's right, how beneficial is the feedback system when a yellow star costs you less than a dollar?[5]

## The Limits of Policing

Fraud has long been a concern on eBay. According to a 2006 report by the International Data Corporation (IDC), over half of Microsoft-branded software offered on eBay was illegitimate.[6] More recently, in August 2010, a rash of fake Apple iPads was sold on the site.[7] Other scams don't even require making a sale. For instance, under the so-called "eBay second chance scam," bidders who lose an auction receive an official-looking email that appears to be from eBay, stating that the seller has given them a second chance to win the item. (eBay does in fact allow sellers to offer second chances, but notifications always come through eBay's internal messaging system, called eBay Messages.) Buyers who click on the link, and then enter their personal details, will have their credit card and/or identity stolen.

eBay could use its policing powers to eradicate at least some scams. For instance, consider the second chance scam. In 2010, SafeFrom Scams.com reported that "hundreds of thousands of [second chance scam] e-mails are sent every month," making this a significant threat. Public reports of this scam have persisted for years, at least since 2004.[8] The second chance scam leverages the fact that eBay allows sellers to communicate with bidders. This is done with good intentions, to allow them to share information about the item for sale. After the auction, however, there's no good reason for a seller to communicate with bidders outside of eBay's official channels. Sellers who do so are either looking to take advantage of bidders (by stealing their identities or squeezing a bit more money out of them) or to take advantage of eBay (by avoiding seller fees).

Fortunately, eBay can stop illicit second chance offers simply by "deputizing" its users. For instance, suppose that eBay were to send the following message to users the first time that they come in second place in an auction:

Thank you for participating in an eBay auction. We hope you agree that bidding on eBay is a convenient (and even fun!) way to find good products at a great price, even when you don't win. As someone who came in second place in an auction, we want to alert you to a tactic that a few disreputable sellers are using on the site— until we catch them and kick them out! What they'll do is send an email (not through eBay Messages) giving you a "second chance" to win. Any such email is illegitimate, and sending one is immediate grounds for account suspension. If you receive such an email, do not reply. Simply forward it to us and we'll take care of the rest. With your help, we can keep eBay the safe and secure place that we hope you've come to love as much as we do. Happy bidding!

It's puzzling that eBay hasn't already pursued such an approach. Perhaps eBay is simply hesitant to highlight the fact that fraud happens on the site, for fear of scaring away users and/or generating bad press. However, by giving buyers clear steps to take should they be approached by a second chance scammer, and by implicitly guaranteeing that eBay will "get the bad guy," they could actually build confidence that eBay is taking a proactive stance to protect users against fraud. Indeed, news stories reporting on such an effort could actually enhance the brand.

Of course, eBay does actively seek to catch scammers and cooperates with law enforcement to put the worst ones behind bars. For instance, in 2010, an Arizona man was sentenced to twenty-one months in prison for selling hundreds of thousands of dollars worth of pirated software on eBay, while a Texan was sentenced to thirty months for fleecing $191,000 from forty-six customers in a "hot tub scam" in which he never delivered the promised tubs.[9] In other cases where the evidence of crime is not as clear-cut, the most that eBay can do is suspend the suspect's account. For instance, in July 2011, "BruntDog" shared his frustration at Strat-Talk.com, a forum for those who love Stratocaster guitars, over a recent purchase of a guitar that had been "gutted . . . the only Fender parts on it were the neck, tuners and case." He managed to get his money back but, lo and behold, the same guitar surfaced again:

Not only didn't he disclose the fact that this isn't a Stratocaster, he relisted it with the very same listing that he used to scam me!! So fellow Strat-Talkers, avoid this seller like the plague. I am going to report the item and seller to eBay as soon as I finish this post. Anyone else that wants to do the same, it would be greatly appreciated. I'm thinkin' a little piling on is called for here. JFYI, this seller was made very aware of the changes made to this guitar and ADMITTED that it was not a Strat. I have all of the correspondence and there is absolutely NO WAY this guy isn't trying to pull a scam.[10]

As it turns out, BruntDog was not successful at getting this seller off eBay. In fact, when I checked in October 2012, this seller had an excellent eBay reputation score, with 99.6 percent positive feedback over the past twelve months. Interestingly, though, his few negative feedbacks were suspiciously reminiscent of BruntDog's complaint. For instance, one buyer reported, "Guitar came in a bizarre condition. He hides it on the photos. Really nightmare," to which he replied, "LIAR! Claimed guitar was refinished and wanted a $100 refund. LIES! Retaliation!" Another complained, "Wrong neck on Bass, Totally un-playable and no shipping costs refunded by seller," to which he replied, "Completely inaccurate, bass is like new, never altered. Buyer is trying to scam."

What if BruntDog had been able to convince eBay to boot this seller off the site? Even that wouldn't have been the end of the story, since scammers can always open new accounts and try again. Indeed, dozens of valid eBay usernames are available for sale at the eBay Suspensions Forum, a site catering to people who have been suspended from eBay. In 2012, for example, $250 could have bought you "A fully verified USA eBay account with increased selling limits (100/$5,000)" that is "ready to sell" with "at least 2 buyer feedbacks."[11] Given the ease with which cheaters can get back on eBay, policing by itself is clearly not enough.

## Twin Threats: Seller Fraud and Buyer Extortion

Before we can craft solutions to restore buyer trust on eBay, we need to dig deeper into the source of the problem. After all, brick-and-

mortar retailers rarely succeed with a business model of intentionally selling deficient products, or of not delivering the goods at all.
It's easy to see why. Imagine that you are buying a DVD from a video
store but that, after you pay, the clerk swaps out your pristine disk for
a scratched one. Seeing this fraud firsthand, you'd get mad and cause a
fuss, disrupting business in the store, and then tell all your friends to
go elsewhere.

eBay's vision of its marketplace is much the same, with negative
feedback serving as the "fuss" that encourages sellers not to cheat.
That said, there are at least two key differences that make life easier
for cheaters on eBay than in a traditional retail setting. First, buyers
don't know where to find eBay cheats in person, so they can't confront
unscrupulous sellers or call in the police. Further, given the nature
of long-distance commerce, some misunderstandings are inevitable.
And such honest misunderstandings provide cover for dishonest sellers who intentionally misrepresent the quality of their goods.

In the stamp-collecting world, for instance, collectors can disagree in good faith whether a stamp is (say) "fine" or "very fine." If
a buyer is able to inspect the stamp in person, such disagreements
are irrelevant. The buyer will form an opinion and only purchase
the stamp if the price is right. When collectible stamps are sold over
the Internet, however, even a legitimate seller will sometimes field
complaints from buyers who think that the stamp is "not as advertised." Some unscrupulous buyers use such complaints as a means to
force sellers into accepting lower prices, with the threat of negative
feedback if they do not. Such "buyer extortion" even makes it possible that unscrupulous sellers may wind up with fewer negative feedbacks than upstanding sellers.

Here's how that might work. Consider first a perfectly honest stamp
dealer who always rates her stamps correctly and provides impeccable
service. Such a seller will rarely get complaints from legitimate buyers, but will provoke negative feedback from buyer-extorters when
she refuses to unfairly lower the price. Consider next an unscrupulous dealer who consistently rates "fine" stamps as being "very fine,"

but who readily resolves any dispute over quality by accepting a lower price. Buyers who lack the expertise to assess quality will not complain, despite the fact that they are paying an unfair premium. More expert buyers (as well as extorters) will complain but, as long as the unscrupulous seller defers to their judgment and takes a lower price, they likely won't submit negative feedback. In this way, overall, the unscrupulous dealer could wind up with a better eBay reputation.

Interestingly, the problem of buyer extortion on eBay may have grown after 2008, as an unintended consequence of changes that eBay introduced to address other concerns. Until then, eBay had allowed sellers to rate buyers, as well as buyers to rate sellers. Such two-sided feedback naturally encouraged buyers and sellers to retaliate against one another, hitting anyone who gave them negative feedback with negative feedback of their own. Buyers complained that such retaliation discouraged them from providing honest feedback after a bad experience, and that this lack of honest feedback was allowing unscrupulous sellers to thrive on the site. Acknowledging such concerns, eBay disabled sellers' ability to leave negative feedback in May 2008.

Buyers mostly cheered the move, one commenting approvingly to the *Internet Patrol* website: "Good for eBay.... Finally a smart move on their part. I have 4 negative feedbacks in my entire history. All left in retaliation for me trying to give honest feedback." However, the threat of retaliation also offered some protection for sellers against buyer-extorters. After all, sellers could always defuse a buyer's extortion threat with one of their own: "Go ahead and give me negative feedback. I'll give you a negative as well and, since I'll never give in, we'll just wind up mutually retracting our negatives. So why bother? Just leave me alone and go scam someone else."[12] Not surprisingly, sellers were dismayed when they lost the ability to defend themselves in this way. One seller lamented:

> This move is horrible for honest sellers. There is no shortage of
> unethical buyers out there who will make false damage claims,

or state that the wrong item was sent, an item was missing, etc. With the new policy, unethical buyers can blackmail the seller into getting what they want.[13]

## Restoring Buyers' Trust in Sellers

To protect buyers against seller fraud, eBay promises anyone who uses eBay's PayPal service that "If you don't receive the item, or if it is not as described, eBay Buyer Protection has got you covered." Such protection undoubtedly helps reassure many buyers who might otherwise not use eBay due to fears of seller fraud. However, eBay Buyer Protection does not completely solve the underlying problem of seller fraud.

For one thing, reimbursement is not assured if the seller disputes a buyer's claim. In that case, the buyer must establish to eBay's satisfaction that the item is "not as described," but this may be easier said than done. For instance, suppose that a buyer purchases a painting that turns out to be a forgery. The buyer will need to find an expert to verify that the painting is fake. Should the seller provide contradictory "expert testimony" (falsely) vouching for its authenticity, eBay might not be able to judge who is telling the truth. Moreover, eBay gives the seller the right to "refuse the return" if the seller claims that the product has been damaged since its initial shipment. If so, eBay may only partially reimburse the buyer.

A more fundamental issue is that, since eBay Buyer Protection covers all sellers who accept PayPal, it doesn't do much to help buyers differentiate between trustworthy sellers and frauds. Indeed, to the extent that Buyer Protection lulls buyers into a sense of security, it could actually lead to *more* fraud on the site, as fraudsters find it easier to lure buyers into purchasing from them. (That said, Buyer Protection also empowers eBay with more tools to catch frauds, since processing more payments through PayPal gives eBay access to data that can potentially be used to spot cheats.)

## IDEA: HELP HONEST SELLERS SIGNAL THEIR TRUSTWORTHINESS

The appeal of Buyer Protection is that it protects buyers *after* a fraud has occurred. Even better would be to protect buyers *before* fraud can occur, by steering them to trustworthy sellers. eBay reputation scores are intended to help buyers in this way but, as we have seen, crooks can also maintain excellent reputation scores. Thus, an excellent eBay reputation doesn't necessarily provide a strong or credible "signal" of a seller's trustworthiness.

In game theory, "signaling" means conveying information about yourself to others through the choices you make. For example, quitting your job and moving to remain close to your significant other is a clear signal that you are serious about the relationship, since someone who isn't serious would never make that sort of sacrifice. Recognizing this, leaving town might actually be a smart move for your significant other, since it forces you to make a choice that will reveal whether you are serious. In much the same way, giving sellers more options could be a smart move for eBay, if doing so allows honest sellers to signal their trustworthiness by making choices that frauds would never make.

Sellers' inability to prove that they can be trusted is an especially severe problem for new sellers, who all start with a reputation score of zero. As long as buyers avoid new sellers, fearing fraud, it can be very difficult for honest sellers to attract enough customers to earn a good reputation. Google's Michael Schwarz has a clever, patent-pending idea to break this chicken-and-egg problem, by giving new sellers the option to boost their initial reputation score by *prepaying commissions* to eBay.[14] The seller can get that money back later under this "Schwarz System," but only if he receives enough positive feedback.

A key feature of the Schwarz System is that reputation is measured in dollars[15] rather than with a simple count of positive and negative transactions—specifically, Reputation = (Total *Commissions Paid to eBay* on Positive Transactions) – (Total *Transaction Value* on Negative Transactions). Furthermore, sellers may at any time choose to boost their reputation scores by prepaying commissions to eBay.

For example, a new seller who prepays $100 to eBay will start with a reputation score of 100. Suppose now that this seller conducts a $100 transaction that carries a $5 commission. If the seller receives positive feedback, eBay will not charge any commission, the seller's reputation score will remain at 100, and the seller's bank of prepaid commissions will shrink to $95.[16] If the seller receives negative feedback, eBay will reduce his reputation by the full value of the transaction, all the way from 100 to 0, thereby effectively voiding his $100 prepayment.

Whereas "a yellow star [can be had by crooks] for less than a dollar" under eBay's traditional reputation system, under the Schwarz System it always costs $100 to build up your reputation by 100 points. By design, then, any crook who commits fraud on a $100 transaction is likely to lose $100 worth of reputation, as the cheated buyer will submit negative feedback that knocks down the crook's reputation score by 100 points. The beauty of this idea is that crooks no longer have an incentive to build up their reputation in order to defraud someone later, since the act of committing fraud is itself no longer profitable. Understanding this, sellers who would like to commit fraud will choose not to do so, or simply stay away from eBay.

The main drawback of the Schwarz System is that it presumes a properly functioning feedback system. If buyers never give negative feedback, or if honest sellers are as likely to get negative feedback as dishonest ones, reputation scores will continue to have little signaling value about who can be trusted.[17] Moreover, by increasing the monetary value of buyer feedback, the Schwarz System itself could undermine the integrity of feedback by exacerbating the problem of buyer extortion. In particular, knowing that negative feedback will cost the seller the full value of the transaction, unscrupulous buyers may be emboldened to (say) demand partial refunds for "damage" that does not exist.[18]

Fortunately, there are many other ways by which eBay can empower honest sellers to signal their trustworthiness, without exacerbating the problem of buyer extortion. For instance, suppose that eBay were to launch a program that I shall call "Extra Assurance." Under this program, sellers have the option to grant buyers the right to return their merchandise for a certain number of days after receipt, for any reason whatsoever, for a full refund of the purchase price *less* two-way shipping costs.[19] (To firm up such a guarantee, the money to issue such refunds could be held in escrow by PayPal and only released to the seller after the buyer has not exercised the right to return the item.)

By design, sellers offering Extra Assurance never gain from sending defective merchandise. To see why, consider what would happen if a buyer with Extra Assurance received a defective item. To receive a full refund *plus* free shipping, the buyer might first file a claim under eBay's standard Buyer Protection. The seller could potentially dispute the claim to block such a refund, but the buyer would still have the right to claim a no-questions-asked refund through the Extra Assurance program. Either way, we would expect buyers always to return defective merchandise, so that crooks themselves never get any money from fraudulent sales.

Since sellers who intend to commit fraud have little hope of earning a profit if they offer Extra Assurance, buyers can safely conclude that any seller who does offer Extra Assurance intends to conduct an honest transaction—regardless of that seller's reputation score. Indeed, once buyers come to understand the *strategic* significance of Extra Assurance, they might demand Extra Assurance of any seller before doing business with them. Such a de facto requirement would be devastating to crooks, who would find eBay such an unattractive place to commit fraud that most might simply leave the site.

If Extra Assurance would be so good at rooting out seller fraud, why not mandate that all sellers offer Extra Assurance?[20] To see why not, consider the problem that some stamp sellers would face if forced to offer a no-questions-asked return option. A common practice in the stamp-collecting world is to sell stamps by the kilogram in unsorted batches (called "kiloware"). A dishonest buyer could replace all valuable stamps in the batch with worthless ones and then return the batch for a fraudulent refund. Such "return fraud" is very difficult to catch and punish.[21] Consequently, we can expect return fraud to flourish should eBay ever mandate Extra Assurance for kiloware sales, perhaps even to the point of shutting down the entire eBay kiloware market.

At the same time, being able to offer Extra Assurance could be helpful for sellers of valuable stamps. Seller fraud is a serious concern among stamp collectors on eBay, so honest sellers have a genuine need to signal their trustworthiness. As *Stamp News Australasia* writer Glen Stephens noted in June 2011: "eBay is a terrific resource for stamp collecting. It allows elusive stamps, topicals/thematics especially, to be sourced for a few dollars from a global marketplace.... Sadly the cons find it just as handy, and we owe it to the hobby to keep the cons OUT and the market clean."[22] Furthermore, sellers of high-quality stamps could (in principle) document exactly what stamps they send, allowing them to prove return fraud when it is attempted. The risk of such discovery might deter dishonest buyers from attempting the fraud, in which case offering Extra Assurance could be an essentially costless option for honest sellers. If so, we would expect Extra Assurance to be widely adopted for high-quality stamp sales, driving sales up for honest sellers while running the crooks out of this market.

## Restoring Sellers' Trust in Buyers

Seller fraud gets the most press, but buyer extortion is potentially a more pernicious threat. If buyer extortion becomes too common, honest sellers will simply exit eBay and sell their goods through other channels, so that all honest buyers suffer as well. That's also true of seller fraud—if it becomes too common, honest buyers will stop coming to eBay, hurting all honest sellers—but there is a crucial difference. The most egregious sorts of seller fraud are relatively easy to prove, since the cheated buyer has possession of the falsely advertised product. For instance, when BruntDog the guitar lover was fleeced, he had the phony Stratocaster in his possession. If his alleged scammer had not offered a full refund, BruntDog could have easily escalated the situation, perhaps even triggering legal action.

By contrast, buyer scams can be essentially impossible for an honest seller to prove. For example, an eBay listing for an "Antique Belgian 8 Day Grandfather Clock Circa 1787" in "perfect working order" with a Buy It Now price of $3,900 recently caught my eye.[23] An unscrupulous buyer could try to extort the seller of this clock into taking a lower price through the following sort of false-damage scheme: Once the clock arrives, the buyer opens it up and removes several essential parts. The buyer then takes it to a professional clock restorer who (i) documents that it is not in working order and (ii) provides a written estimate that it will cost (say) $1,000 to fix the clock. The buyer then posts negative feedback and, moreover, initiates a formal claim of seller fraud with eBay, demanding a $1,000 refund to reflect how much it will cost to get the clock ticking again. (Once that money is in hand, of course, the buyer in this story won't actually fix the clock but simply return the missing pieces and keep the cash.)

If the seller cannot irrefutably prove that the clock wasn't missing any pieces when it was shipped, eBay may be unwilling to intervene to remove the buyer's false negative feedback. If so, the only way for the seller to restore her reputation is to give in and pay the extorter. Moreover, since the seller cannot post negative feedback of her own, the buyer faces no personal risk when extorting a seller in this way.

### IDEA: ELIMINATE USERS' ABILITY TO RETALIATE
A fundamental feature of eBay feedback is that it's changeable, i.e., that buyers can remove or edit their negative feedback. But why? The purpose of eBay feedback, as originally envisioned by eBay founder Pierre Omidyar, was to give buyers and sellers an opportunity to communicate their experience with others. ("Make your complaints

in the open. Better yet, give your praise in the open. Let everyone know what a joy it was to deal with someone.")

The problem is that users have hijacked the feedback system to influence the transaction process itself, threatening negative feedback unless they get what they want. This wouldn't be so bad if eBay users were all bound by a strict moral code, as the threat of negative feedback would then typically be justified, with appropriate demands. However, as we've seen, unscrupulous users don't settle for what's reasonable or right. Unscrupulous buyers use the threat of negative feedback to extract unfair benefits, while unscrupulous sellers (when given the chance to offer negative feedback of their own) seek to squelch honest buyer feedback.

Fortunately, there is a simple solution: restore sellers' ability to offer negative feedback about buyers, but transform feedback into a "simultaneous move" game. That is, force both parties to a transaction to submit their feedback without knowing what feedback the other side is giving. Here's one way that might work.

### Step 1: Transaction and private communication
As long as the transaction remains "open," the buyer and seller may communicate privately via eBay Messages to raise and resolve any concerns. However, they may not submit feedback until the transaction is "closed."[24]

### Step 2: Simultaneous public feedback
Once the transaction is "closed," both buyer and seller have a fixed amount of time (say, one week) to submit feedback. Both sides' feedback is then publicly posted, at which point it becomes irrevocable and unchangeable.

### Step 3: Simultaneous public response to feedback
The buyer and seller then each have the opportunity to post a public response to any feedback posted about them. Like the feedback itself, these responses are collected but not revealed until a preappointed

time (say, one more week), at which point they are publicly and irrevocably posted online.

The two key features here are (i) two-sided feedback and (ii) simultaneous feedback.[25] Allowing sellers to give feedback about buyers is essential, to make buyer extortion less attractive (and help eBay identify extorters). For instance, consider again the unscrupulous clock collector who falsely claims that a clock is not in working order. Under a simultaneous feedback system, the dealer might still feel compelled to give in to the extorter, in order to minimize the risk of getting negative feedback. However, because feedback is given simultaneously and cannot be changed later, the dealer can now safely offer negative feedback ("This buyer scammed me") without the risk of retaliation in the feedback itself.[26] Moreover, since all feedback is submitted after the transaction is closed, a seller gains no benefit from making an accusation of extortion, giving such accusations an automatic measure of credibility.[27]

In much the same way, simultaneous feedback frees buyers to share candid negative feedback about sellers, as sellers cannot hit buyers with retaliatory negative feedback of their own. So, simultaneous two-sided feedback can help protect buyers from seller fraud even as it protects sellers from buyer extortion.

## Fulfilling the eBay Vision

Pierre Omidyar founded eBay with a vision "to make it easier to conduct business with strangers over the net." eBay can fulfill that vision even more fully by taking additional steps to protect buyers from seller fraud and to protect sellers from buyer extortion. Given the depth and complexity of eBay's marketplace, finding a solution might seem hopeless, especially since the crooks out there will always be looking for a way to get around whatever eBay tries. However, viewing the eBay community through the game-theory lens allows one to identify how and why dishonest buyers and sellers are able to thrive on the site. Ultimately, the weakness of eBay's feedback and reputation system can be traced to two key strategic factors:

1. *Asymmetric information*: eBay buyers can't tell which sellers are frauds (even an excellent reputation score isn't enough to prove that you are trustworthy), while eBay sellers can't tell which buyers will try to extort them.
2. *Intimidatory tactics*: Having the ability to post and modify negative feedback allows unscrupulous buyers to extort honest sellers.

Solving the asymmetric information problem for buyers requires either (i) ensuring that sellers' reputation scores truly reflect their quality or (ii) finding some way other than reputation scores by which sellers can signal their quality. For example, offering Extra Assurance (described earlier) could be one way for sellers to credibly signal their intention of conducting a fraud-free transaction, since anyone who offers Extra Assurance and then commits fraud cannot earn a profit.

What about intimidatory tactics? As long as eBay users are capable of offering feedback, unscrupulous users will be able to threaten to give negative feedback. That said, eBay can eliminate their ability to threaten to *retaliate* against honest negative feedback by utilizing a "simultaneous feedback" system in which each party to a transaction cannot observe what feedback the other side has submitted until after submitting their own feedback. Without the ability to retaliate against their honest victims, criminals posing as buyers or sellers on eBay can expect to receive more honest negative feedback, allowing eBay to target and prosecute them more effectively than before. Facing a greater risk of being found out for their fraudulent activities, crooks might even decide that eBay isn't a profitable place to operate and leave the site.

# Antibiotic Resistance

Bath & Body Works touts itself as "a 21st-century apothecary integrating health, beauty, and well-being" that "reinvented the personal care industry with the introduction of fragrant flavorful indulgences." Indeed, they offer forty(!) varieties of liquid hand soap alone, in flavors ranging from "Fresh Picked Tangerines" to "Caribbean Escape" and even "Twilight Woods for Men." Bath & Body Works has maintained a bit of mystery about these soaps on its website, listing their ingredients in July 2012 simply as "Water, Fragrance, Honey Extract." But there is another ingredient in all of these soaps: triclosan, a pesticide and antibacterial antiseptic agent. Including triclosan allows Bath & Body Works to label its liquid hand soaps "antibacterial."

The appeal of antibacterial soap is the much advertised notion that it is "clinically proven" to kill 99 percent of germs. Still, what about the other 1 percent? Presumably, they will survive and multiply. If enough people use antibacterial soap, it is therefore natural to expect the bacteria that remain to develop antibacterial resistance, to the point that everyone could be worse off than if antibacterials had never been added at all.

Of course, the fact that society at large might suffer isn't a compelling reason for individuals to steer clear of antibacterial soap. As long as antibacterial soap provides any added protection, individual consumers have an incentive to buy it. In this way, consumers appear trapped in a Prisoners' Dilemma. Each has a dominant strategy to buy antibacterial soap, but everyone is worse off when everyone uses it due to rising antibacterial resistance.

## Changing the Antibacterial Soap Game

In 2007, *Clinical Infectious Diseases* published a survey of the scientific literature on antibacterial soap that documented "evidence of triclosan-adapted cross-resistance to antibiotics among different species of bacteria."[1] Worse yet, antibacterial soap is "no more effective than plain soap at preventing infectious illness symptoms and reducing bacterial levels on the hands." Indeed, for such soap to have any antibacterial effect, it must sit on the hands for at least two minutes—and no one leaves soap on their hands that long. Consequently, no one is getting any antibacterial benefit from having triclosan in their hand soap.

The impotence of antibacterial soap might seem like bad news, but actually it means that consumers aren't truly locked in a Prisoners' Dilemma. If only consumers knew that antibacterial soap provides no extra protection against bacteria, they would no longer have an incentive to use it. In fact, once consumers learn of the potential health risks of triclosan exposure, they will actually have a dominant strategy *not* to use antibacterial soap. Long regulated by the Environmental Protection Agency as a pesticide, triclosan has been shown to interfere with the endocrine system of animals such as frogs and rats.[2] Moreover, the Centers for Disease Control and Prevention's National Biomonitoring Program has detected triclosan in the urine of 75 percent of Americans (six years and older) sampled.[3]

This scientific evidence, plus sustained pressure from environmental activists, has had a significant policy impact. In May 2012, Canadian regulators declared triclosan "toxic to the environment," a move that will sharply curtail its use in Canada. In the United States, however, the Food and Drug Administration (FDA)'s position is that "Triclosan is not currently known to be hazardous to humans. But several scientific studies have come out since the last time FDA reviewed this ingredient that merit further review."[4]

Would banning triclosan in the United States solve the problem here? Unfortunately, it would not. A ban would take triclosan off the shelves, true, but consumer demand for antibacterial soap would remain. If anything, consumers might feel more confident than before in the safety of whatever replaces triclosan. But that new ingredient will likely be less well studied and potentially even more dangerous. After all, since FDA regulators only ban products that are *known* to be dangerous, firms have a strong incentive to use products that no one knows anything about.

Fundamentally, the problem with today's consumer-safety regulatory environment is that it amounts to a glorified game of whack-a-mole, with our regulatory agencies in the role of the hapless player who misses time and again, as he always aims where the mole used to be.[5] Fortunately, viewing this problem through the game-theory lens reveals ways to change the game and transform consumer protection regulation for the better.

Before we can craft such solutions, however, it's essential to understand the problem more deeply. Why do some firms include antibacterial agents in their soaps, despite the fact that they offer no real protection against bacteria and might even create new health risks? The reason is simple: their customers want it. For decades, consumers have been taught a false doctrine, that bacteria are the enemy and must be destroyed. By offering an "antibacterial" product line, a company like Bath & Body Works is simply responding to this demand, in order to maximize profits. Banning triclosan won't change that. The only real solution is to change what consumers want.

First, regulators could change how products are labeled. Under current practice, including an antibacterial agent like triclosan allows a firm to label its product "antibacterial." However, tossing triclosan into a product doesn't necessarily protect against bacteria, just as slapping wings onto something doesn't necessarily make it fly. What matters is how the product will be used. In the case of hand-washing, even the Mayo Clinic only recommends that people wash for twenty seconds,[6] whereas triclosan needs to sit on the hands for at least two minutes to have any antibacterial effect. What this means is that triclosan-laden hand soap is meant to be washed away before it can have any antibacterial effect. Since such soap offers no actual protection from bacteria, labeling it "antibacterial" is confusing and even misleading. Regulatory agencies such as the FDA and the Federal Trade Commission could therefore reasonably step in to regulate soap and other so-called "antibacterial" products that fail to protect against bacteria.

Second, a credible third party could educate consumers on which products offer true bacterial protection. Such information campaigns have been successful in other contexts. For instance, a decades-long shift in milk farming methods has led to an increased reliance on pesticides, growth hormones, and antibiotics. In 2011 alone, 29.9 million pounds of antimicrobial drugs were administered to livestock in the United States, about four times the amount taken by people,[7] in large part to speed the growth of healthy animals. This widespread overuse of drugs on livestock has created the conditions for a seemingly endless stream of threatening new superbugs, most recently a resistant strain of *E. coli* that has "put 8 million women at risk of difficult-to-treat bladder infections."[8] Worse yet, because all bacteria exchange DNA,[9] antibiotic resistance developed in *any* bacterial strain—even one that itself poses no harm to humans—can eventually find its way into people.

Responding to consumer concerns over such practices, the Department of Agriculture (USDA) created a new category of "organic milk," defined as milk that comes from cows that have been exclusively fed organic feed, have not been treated with synthetic hormones, and are not given certain antibiotics. Organic milk is now a supermarket staple. A "Safe Biotic" designation for products that foster a healthy human biome could, in much the same way, inform consumers and stoke demand for products with a healthful biotic effect on individuals and society at large.

A "Safe Biotic" designation would also give firms an incentive to include only the most biome-healthful ingredients in their products. Even Bath & Body Works might finally[10] follow the principled lead of firms like Colgate Palmolive, whose Softsoap line has been antibacterial-free since January 2011, and Johnson & Johnson, who "set a goal to phase out triclosan in our beauty and baby care products," which include Aveeno, Neutrogena, and Lubriderm, in August 2012.[11]

## The Broader Battle Against Bacteria

> Some experts say we are moving back to the pre-antibiotic
> era. No. This will be a post-antibiotic era.... A post-
> antibiotic era means, in effect, an end to modern medicine
> as we know it. Things as common as strep throat or a
> child's scratched knee could once again kill.
>
> —*Margaret Chan, Director-General of the World Health
> Organization, 2012*

Resistance to antibacterial soap is perhaps the most benign example of a disturbing worldwide trend toward resistance to the antibiotics used to treat most bacterial diseases. Consider tuberculosis. In 1800, nearly 25 percent of all deaths in Europe were caused by this so-called "wasting disease," also known as "consumption" because of the way in which the untreated disease seems to literally consume the living. A truly terrifying illness, tuberculosis may even have inspired the vampire legend. Shortly after a rash of deaths from tuberculosis, others in the community would often also begin to waste away from the disease. Villagers would blame the recently deceased, believing that they had risen from the grave to feed upon the living. It didn't help that, when the dead were dug up, villagers would often find blood draining from their mouths.[12]

Tuberculosis the disease lost its bite in 1946, when scientists discovered that the bacterium that causes it, *Mycobacterium tuberculosis*, was vulnerable to attack by an antibiotic produced by *Streptomyces sp.*, another bacterium. Unfortunately, misuse of this and other antibiotics has led to resistance, to the point that tuberculosis strains resistant to multiple medicines are becoming more common. Indeed, news has come of strains in India that may be "totally resistant,"[*] i.e., resistant to all known antibiotics.[13]

This is especially bad news, since reversing antibiotic resistance is very difficult. As Dr. Dan Andersson, fellow of the American Academy of Microbiology, explains: "Resistance might be reversible, provided antibiotic use is reduced. However, several processes act to stabilize resistance,[14] including compensatory evolution [reducing the disadvantages associated with resistance] ... and genetic linkage or coselection between the resistance markers and other selected markers [making it costly to lose resistance as then the bacteria would lose other benefits]."[15] In particular, should total resistance ever "stabilize" in the sense described by Dr. Andersson, totally resistant bacterial strains may never evolve back to a state that is susceptible to antibiotics—even if doctors everywhere were to cease all antibiotic treatment.

Given this grim prognosis, it's no surprise that scientists are working overtime to identify new treatment strategies. For instance, one

---

*As of December 2012, the World Health Organization has not designated the new Indian strains as totally resistant, because their resistance to some drugs has not yet been established. That said, they are known to be resistant to "the two key and most potent anti-TB drugs [called] isoniazid and rifampin [as well as] to the most potent second-line drugs [including] fluoroquinolones," but "tests against [several drugs including] cycloserine and ethionamide are unreliable" (personal communication from Dr. Mario Raviglione, director of the WHO's Stop TB Department and fellow of the Royal Academy of Physicians).

idea is to recruit viruses called bacteriophages to selectively kill only the bacteria that are resistant to antibiotics.[16] Yet bacteria can develop resistance to viruses as well, potentially limiting the promise of even this revolutionary approach. Indeed, because bacteria are so good at developing strategies to resist just about any assault, many in the medical profession seem resigned to the inevitability of antibiotic resistance. But there is hope. Indeed, recent technological advances in genetic testing have created new strategic options for our game with disease that hold the potential to reverse antibiotic resistance and, in doing so, to tame bacterial disease—forever.

*Recent advances in genetic testing have created new strategic options to put resistant strains at a disadvantage and thereby reverse the frightening trend toward antibiotic resistance.*

The first and most important step toward finding the cure to antibiotic resistance is to change the way we think about disease, viewing it through the game-theory lens. We typically think of disease as a contest between people and the diseases that afflict us (e.g., "she's fighting the flu"), but this view misses an essential element of the game. Disease is a deadly contest, yes, but the fiercest battle is between strains of the same disease, each striving for supremacy in the overall population of bacteria that cause that disease. Each strain's success— whether it dominates the population or dwindles into extinction—is determined by how well it plays three related games:

1. *The Infection Game*: Can the strain get past the human immune system? (Success in this game is called "infectivity.")
2. *The Transmission Game*: Can the strain transmit itself to new hosts? (Success in this game is called "transmissibility.")
3. *The Treatment Game*: Can the strain survive medical treatment well enough to continue transmitting itself? (Success in this game is called "resistance.")

FIGURE 28    The games of disease

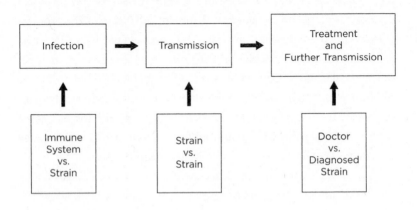

The strategic logic of rising antibiotic resistance is plain to see in figure 28. Suppose that two strains are equally infective (in the Infection Game) and equally transmissible (in the Transmission Game), but that only one strain is resistant to antibiotics (in the Treatment Game). That resistant strain will be more likely to survive treatment, giving it an overall advantage in the games of disease. We may therefore expect the resistant strain to grow more quickly and eventually dominate the bacterial population. (Of course, resistant strains need not always win in the end. If a susceptible strain is more infective and/or more transmissible than a resistant strain, the susceptible strain may still outcompete the resistant one.)

Figure 28 also points to ways to slow or even reverse the rise of antibiotic resistance, if only one (or all) of these games can be changed to put resistant strains at an overall disadvantage relative to their susceptible competition.

# Reversing Unstable Resistance:
## Changing the Infection Game

The human immune system is our first line of defense against disease, and our only defense against totally resistant disease. Recognizing this, the Interagency Task Force on Antimicrobial Resistance (ITFAR)—a collaboration of the Centers for Disease Control and Prevention, the Food and Drug Administration, the National Institutes of Health, and nine other federal agencies—is working to "facilitate development of vaccines for resistant pathogens such as *Staphylococcus aureus, Mycobacterium tuberculosis, C. difficile*, enteric pathogens, and *Neisseria gonorrhoeae*."[17]

Beyond the direct protection they offer, vaccines may also help to reverse antibiotic resistance by reducing the advantage that resistant strains enjoy in the treatment phase. To see why, suppose that a vaccine is developed that offers protection against all strains of a disease.[18] By strengthening the human immune system, such a vaccine may cause many infected patients not to need drug treatment to fight the disease. As fewer drugs are prescribed, resistant strains will then enjoy less of an advantage in the Treatment Game.[19] If those strains' resistance is "unstable" in the sense discussed earlier—i.e., if resistant strains are less infective and/or less transmissible than susceptible strains—they will be at an overall disadvantage. Whenever resistance is unstable, we may therefore expect resistant strains to dwindle in number and perhaps even "voluntarily" shed their resistance by evolving back to their original susceptible form.

That said, vaccines are clearly not a panacea. To push back against resistance in a meaningful way, vaccines must be administered globally. As long as any part of the world is unprotected from a resistant strain, that strain's resistance may stabilize at any time, at which point it could become an essentially permanent threat to humanity. And as the list of stably-resistant pathogens grows over time, so does the list of diseases against which vaccine is our only effective protection. If the number of these stably-resistant diseases ever becomes too great, they may simply overwhelm our ability to protect against them all, like barbarians at the gate.

So, while vaccines will always be an important frontline weapon in our fight against disease, we need to look elsewhere for a winning long-term strategy against antibiotic resistance. Fortunately, recent developments in the diagnosis and treatment of disease offer other options to reverse antibiotic resistance, even without vaccines.

## Reversing Rare Resistance: Changing the Treatment Game

If a doctor knew that her patient's disease was resistant to one drug but susceptible to another, she would always prescribe the more effective medication. Unfortunately, in practice, doctors must often decide which treatment to prescribe without knowing the susceptibility of a patient's disease. The reason is simple: for bacterial diseases such as tuberculosis, it can take weeks to culture a large enough sample to test for susceptibility. No doctor can wait that long before prescribing treatment. All doctors therefore tend to prescribe the same first-line antibiotics—whatever tends to be most effective for most patients—giving a strategic advantage to whatever strains are most resistant to those drugs.

While done with good intentions, this practice creates the conditions under which resistance to first-line antibiotics can potentially emerge. Of course, should such resistance arise and become widespread, first-line antibiotics won't be as effective anymore. If so, doctors will then naturally move on to the next-best "second-line" drugs, creating the conditions for resistance to those antibiotics to emerge as well. On and on this cycle may go, until no good options remain.[20]

The only way to break this logic is to empower doctors with the tools to diagnose the susceptibility of a patient's disease as quickly as they can diagnose the disease itself. The good news is that, with recent advances in genetic testing, the capability for such quick susceptibility diagnosis is finally at hand. Indeed, the 2nd Infectious Disease World Summit, held in San Francisco in July 2012, was abuzz with talk of a new approach to genetic testing of disease being developed by Cepheid, the molecular diagnostics firm. Rather than waiting to grow a culture of bacteria, Cepheid's GeneXpert system hunts directly in biological samples for targeted strands of DNA. This allows the GeneXpert to determine if a specific bacterial strain is present in any given sample, without needing to isolate or culture the bacteria in question.

This new technology allows doctors, for the first time, to diagnosis the susceptibility of disease (i.e., which drugs will be most effective at combating a patient's disease) in addition to diagnosing the disease itself. In April 2011, the FDA approved Xpert Flu, a diagnostic test that "simultaneously detects and differentiates Influenza A, Influenza B, and the 2009 H1N1 influenza virus in about one hour."[21] What about tuberculosis? The Xpert MTB/RIF test, so named because it detects both the presence of *Mycobacterium tuberculosis*[22] and resistance to the antibiotic rifampin,[23] was introduced in 2009. Moreover, thanks to a fast technology transfer facilitated by the World Health Organization, Xpert MTB/RIF testing capability is already in place in seventy developing and EU countries.[24]

Xpert tests like these will be a powerful new tool for doctors, as they can target drugs more effectively at partially resistant strains of a disease. This helps to level the playing field in the Treatment Game, but does not level it entirely. For one thing, as a for-profit firm, Cepheid will undoubtedly charge for its diagnostic tests, and not everyone may be able to pay, especially in poorer parts of the world. Recognizing the importance of getting Xpert testing into developing countries, a consortium supported by a UNITAID grant to the World Health Organization, including the United States President's Emergency Plan for AIDS Relief (PEPFAR), the United States Agency for International Development (USAID), and the Bill & Melinda Gates

Foundation, announced in August 2012 an agreement with Cepheid to "reduce the cost of Xpert MTB/RIF cartridges from $16.86 to $9.98, a price which will not increase until 2022."[25] This is great news, as more doctors in India, China, and elsewhere are now likely to diagnose resistant strains of tuberculosis more quickly.

That said, the Xpert MTB/RIF test only detects resistance to rifampin, whereas the standard first-line treatment is actually a cocktail of several drugs (rifampin plus isoniazid, pyrazinamide, and ethambutol). Not knowing whether a patient's disease is resistant to these other drugs complicates effective treatment. To see why, imagine that a tuberculosis patient is diagnosed as being rifampin-resistant. As long as the remaining first-line drugs are sufficiently effective for *most* rifampin-resistant patients, doctors will naturally tend to prescribe a cocktail of those other drugs. While effective against strains that are resistant only to rifampin, such a treatment approach allows strains that are resistant to multiple first-line drugs to remain at an advantage. In the end, then, being able to diagnose rifampin resistance may not be enough to halt the overall trend toward simultaneous resistance to many drugs.

To address this problem, it's essential that we develop additional diagnostic tools to detect resistance to other drugs that treat tuberculosis. Unfortunately, there's not much profit motive for a firm like Cepheid to develop such new diagnostics. Cepheid already has a product on the market, the Xpert MTB/RIF test, that is good enough to be widely adopted. And while patients and doctors would undoubtedly welcome an even better test, it's unclear whether they can afford to pay more for it, especially in the poorer parts of the world where tuberculosis is most prevalent. From Cepheid's perspective, then, there may be more bang for the buck from developing tests that drive sales by opening up new diseases to molecular diagnosis, even though humankind might benefit most if Cepheid focused on developing a deeper arsenal against just our very toughest enemies.

For the sake of argument, though, suppose that Cepheid did not face any such constraint and could offer an affordable test that diagnoses susceptibility to all known antibiotics. Would such a complete diagnostic tool, by itself, empower doctors to completely halt the rise of antibiotic resistance? Perhaps not. Yes, doctors would be able to target and kill susceptible and partially resistant strains more effectively than ever before. But what about strains that are "totally resistant," i.e., resistant to all known antibiotics? Absent any effective antibiotic treatment, the only way to stop totally resistant strains from continuing to transmit themselves is to impose physical barriers, as in the following "Detect + Isolate" strategy:

1. *Detect*: Test every patient for resistance using a quick molecular diagnostic tool such as Cepheid's GeneXpert system.
2. *Isolate*: If heightened resistance is detected, isolate the patient until his/her disease is no longer transmissible.

Of course, only complete isolation can ensure that totally resistant strains don't continue transmitting themselves. Should total resistance ever become sufficiently widespread, however, it may become impossible to isolate every patient diagnosed with totally resistant disease.

Overall, then, whether even perfect GeneXpert diagnosis will be enough to stop the rise of totally resistant disease depends on how prevalent it has already become. As long as total resistance is sufficiently rare, it may be feasible to isolate it completely enough to eliminate the advantage that totally resistant strains would otherwise enjoy in the Treatment Game. Indeed, since even susceptible strains are at least somewhat capable of transmitting themselves after treatment, completely isolating all diagnosed cases of totally resistant disease could actually put totally resistant strains at a disadvantage in the Treatment Game relative to their susceptible competition. Thus, there is reason to hope that, if GeneXpert testing expands quickly enough to cover more types of drug resistance before such resistance becomes too prevalent, widespread adoption of the GeneXpert system could help to reverse even total resistance.

*If GeneXpert testing expands quickly enough to cover more
types of drug resistance before such resistance becomes too
prevalent, widespread adoption of the GeneXpert system could
help reverse even total resistance.*

Should total resistance grow prevalent enough to overwhelm the medical infrastructure's capacity to isolate it, however, there's no way through treatment alone to stop totally resistant strains from acquiring a "monopoly" over the disease. Consequently, while Cepheid's GeneXpert diagnostic system is a "game-changer" for the treatment of disease, it may not be enough on its own to defeat antibiotic resistance everywhere in the world. Indeed, by allowing doctors to wipe out susceptible and partially resistant strains so much more effectively than in the past, widespread adoption of the GeneXpert might even make the problem worse—by accelerating the spread of total resistance[26]—perhaps especially in developing countries that lack the capacity to mount an effective isolation program.

Fortunately, the capability to quickly diagnose whether bacteria are susceptible to drug treatment transforms the games of disease in still other ways that could potentially turn the tide and reverse total resistance, even in places where it is too widespread to isolate effectively. In particular, having the capability to diagnose drug susceptibility in a matter of hours, rather than days or weeks, creates new strategic options by which people can influence *transmission* as well as treatment.

## Reversing Widespread Resistance:
## Changing the Transmission Game

When you wash your hands, you are changing the Transmission Game by making it more difficult for all bacteria to transmit themselves. Such transmission-fighting steps decrease the overall burden of disease but, since they are equally effective against all strains of disease, don't favor one strain over another. But what if people who are infected (or at risk of infection) by a resistant strain were more protected than others from transmitting (or receiving) the disease? If so, resistant strains would be put at a disadvantage at the transmission stage, allowing susceptible strains to eventually take over the bacterial population of the disease. This observation motivates the following "Detect + Search" strategy:

1. *Detect*: Test every patient for resistance using a quick molecular diagnostic tool such as Cepheid's GeneXpert system.
2. *Search*: If heightened resistance is detected, launch an epidemiological investigation to identify and test all those (at home, school, work, etc.) who might have contracted the disease from the identified patient, taking steps as well to slow transmission and/or to speed up diagnosis to shrink the transmission window for these people. Should others with resistant disease be found, continue the search among all those who might have contracted the disease from them.

For example, suppose that a school-aged child is diagnosed with a highly resistant strain of some disease. To tip the scales against resistance, one could dispatch a team to that child's school to diagnose infected children even before the disease has progressed to the transmission phase and, if available, to provide antibiotic treatment (called prophylaxis) to protect uninfected children from catching the disease and/or to limit its transmissibility in those who have already been infected.[27]

If all students participated in this "Detect + Search" program, the resistant strain could potentially be stopped cold at the school. However, full participation is not essential.[28] If a portion of the student body opts out of screening and/or prophylaxis, subsequent transmission will still be slower than if no one had participated at all. Thus, as long as susceptible strains do not face the same sort of intensive proactive detection and treatment, these susceptible strains will have an advantage and—slowly but surely—grow as a fraction of the overall bacterial population, perhaps even to the point of putting resistance on the path to eventual extinction.

# The Path to Victory over Carbapenem Resistance

In 1992, some hospitals began to detect cases of carbapenem-resistant *Enterobacteriaceae* (CRE), rod-shaped bacteria (including the famous *E. coli*) that cause many familiar diseases such as salmonella. Carbapenem is an important class of drugs that includes "antibiotics of last resort" for many bacterial infections. As you can imagine, then, CRE can be quite deadly. In fact, these bacteria are probably even deadlier than you imagined, with mortality rates of 40–50 percent. Worse still, CRE are difficult to eradicate, especially in a hospital environment where bacteria are easily spread in shared facilities, on shared equipment, and so on. Moreover, as patients are transferred between hospitals, or between acute and chronic care facilities, CRE have migrated over the years to more and more hospitals, all over the world.

Fortunately, by taking aggressive steps to interrupt CRE transmission, some hospitals have been able to eradicate these bacteria from their facilities. Efforts to contain the spread of CRE have even been successfully scaled to the national level, in Israel.[29] The Centers for Disease Control and Prevention wants to duplicate that success in the United States, and there's every reason to be optimistic. In June 2012, the CDC published guidance for hospitals on how to fight back against CRE, advice that boils down to the following "Detect + Isolate + Search" strategy:[30]

1. *Detect*: Identify cases of CRE in your hospital.
2. *Isolate*: Remove CRE sufferers, once identified, from the general population.[31]

3. *Search*: Test all patients who are at risk of contracting CRE, including all who are "epidemiologically linked" to any identified CRE sufferer.

This strategy to fight CRE combines both of the key resistance-reversing ideas developed earlier in this chapter:* (i) isolating resistant disease, once diagnosed, takes away CRE's advantage in the Treatment Game and (ii) proactively testing and then treating at-risk patients puts CRE at a disadvantage in the Transmission Game.

Of these two tactics, isolation might at first seem to be the most important. After all, removing CRE from the general population means that there is less of the bacteria floating around to infect additional people. That's true. However, the test for CRE is imperfect,[32] so some bacteria will always slip through the cracks. And as long as CRE continue to have a reproductive advantage due to their resistance to antibiotics commonly used in hospitals, even a small population of "loose" CRE can eventually grow into a hospital-wide monster.

Seeking out and testing at-risk patients in the general population changes all that. For instance, patients who previously suffered from CRE are more at risk of developing it again. Testing these people immediately upon arrival at the hospital makes it harder for carbapenem-resistant strains of a disease to invade the hospital from the outside, compared to susceptible strains that aren't subjected to the same sort of initial screening. Similarly, testing all patients who came into contact with known CRE sufferers (roommates prior to isolation, patients who shared a potentially contaminated machine, etc.) makes it harder for carbapenem-resistant strains to transmit

---

*I developed the "Detect + Isolate" and "Detect + Search" strategies to reverse antibiotic resistance in summer 2012, unaware that the CDC had just published guidance to hospitals to eradicate CRE that was, essentially, a combination of these ideas. As CDC Associate Director for Healthcare Associated Infection Prevention Programs Arjun Srinivasan wrote to me later, in the fall: "Hopefully this means that we're both on the right track."

themselves within the hospital, compared to susceptible strains that can jump relatively unmolested from host to host.

As long as CRE-targeted interventions are aggressive enough, we can expect CRE to be at an overall disadvantage relative to susceptible strains, even in the unisolated general population. This disadvantage, if sustained long enough, will cause CRE to dwindle among the hospital-wide bacterial population and, eventually, even go extinct.

# Extensively Drug-Resistant Tuberculosis (XDR-TB): A Tougher Enemy

In 2005, a team of researchers led by Professor Neel Gandhi of the Yale School of Medicine descended on a "resource-limited" rural hospital in KwaZulu–Natal, South Africa, to document the prevalence of drug-resistant tuberculosis.[33] Of 542 patients diagnosed with active tuberculosis, 221 carried "multi-drug resistant" (MDR) strains that were resistant to both isoniazid and rifampin, the two most potent first-line drugs. Moreover, fifty-three of these MDR patients were actually "extensively drug-resistant" (XDR), meaning that their tuberculosis was also resistant to multiple second-line treatments.[34] Sadly, only one XDR-TB patient survived more than one year, the other fifty-two dying after a median survival time of just sixteen days after being identified by Gandhi's team.

The deadliness of this XDR-TB strain is enough to scare anyone, but what's even more frightening is the ease with which it spread inside the hospital. Gandhi and his colleagues wanted to understand the origin of the XDR-TB epidemic that they had observed, so they carefully traced each patient's past contacts. Their conclusion: most of these XDR-TB sufferers, including six health care workers, had been infected at the hospital itself. The good news, if one can call it good news, is that these patients were so ill that they didn't have the opportunity to spread their XDR-TB to other hospitals as well. But it's only a matter of time until another, somewhat less deadly XDR-TB strain emerges that can both (i) spread itself *within* hospitals and (ii) spread itself *across* hospitals as patients seek care in multiple facilities.

To combat the rise of XDR-TB, we can look to our experience and success at fighting carbapenem-resistant *Enterobacteriaceae*. Just as hospitals have been able to control (and even eradicate) CRE using a "Detect + Isolate + Search" strategy, so we may be able to control (and perhaps even eradicate) XDR-TB by rolling out an aggressive program to (i) detect who is carrying XDR-TB, (ii) prevent those patients from spreading their disease after diagnosis, and (iii) quickly test those who are at risk of having contracted XDR-TB from a known sufferer. That said, there are several crucial strategic differences that will likely make XDR-TB a much tougher enemy to defeat than CRE.

## CHALLENGE 1: DIAGNOSING XDR-TB

There still is no easy way to diagnose XDR-TB. While the Xpert MTB/RIF test allows us to quickly detect resistance to rifampin, we remain incapable of quickly detecting resistance to any other TB-fighting antibiotic. Unfortunately, rifampin resistance is already sufficiently widespread that it's not realistic to isolate all rifampin-resistant patients, especially given the limited resources available in the places where TB is most widespread. Thus, the Xpert MTB/RIF test appears unlikely to be enough to enable a successful "Detect + Isolate + Search" strategy against XDR-TB.

To overcome this challenge, we need to develop affordable tests to quickly diagnose resistance to other drugs, in addition to rifampin. Fortunately, though, we don't need to be able to diagnose resistance to *all* antibiotics, just a carefully chosen few. For instance, suppose that a test were developed that detected resistance to rifampin and isoniazid, the two most potent first-line drugs (to which joint resistance is a growing problem), and to fluoroquinolones, the most potent class of second-line drugs (to which resistance is still relatively rare), but not to any other antibiotic.

Such a diagnostic test would divide TB strains into three basic groups: (i) strains that are susceptible to rifampin and/or isoniazid, (ii) strains that are resistant to rifampin and isoniazid but susceptible to fluoroquinolones, and (iii) strains that are resistant to all three. Type (i) can be effectively treated by a cocktail consisting only of first-line drugs, while type (ii) can be effectively treated by second-line drugs. For type (iii), finally, doctors can take steps to isolate the disease (while also trying other drugs to which resistance remains unknown). In every case, doctors can use the results of this three-drug test to identify an effective method with which to treat or at least halt transmission of the disease, without needing to know its susceptibility to all the other drugs that can treat TB.

The biggest potential weakness of this approach is that doctors may not be able to isolate all type (iii) patients, once joint resistance to rifampin, isoniazid, and fluoroquinolones becomes too common. That's why it's essential to include in the mix a diagnostic for a drug, here fluoroquinolones, to which resistance is still fairly rare. That way, the set of patients for whom there is no clear treatment can be effectively isolated without putting too much strain on hospitals and other local medical resources.

#### CHALLENGE 2: GETTING AHEAD OF THE DISEASE

One reason why tuberculosis is so difficult to fight is that it typically enjoys a long "transmission window" before its symptoms prompt medical attention and treatment. What this means is that, when a TB sufferer presents at the hospital, he has likely already had ample opportunity to expose many others to the disease. Not only that, those people may also have already passed the disease on to still more people. This makes it much more difficult to get ahead of the disease with a "Detect + Search" strategy. Indeed, to make a meaningful dent in disease transmission, it may be necessary for the teams that are conducting the search to identify not only those who were infected by a known patient but also those whom those people infected, and so on.

Fortunately, this is not as unrealistic as it sounds. In November

2012, *Science* published an article on how "high-throughput genetic sequencing" can allow forensic epidemiologists to hunt down those whom infected people may have acquired their disease *from*, as well as those whom they may be passing it on to, in effect mapping out the entire epidemiological network of an in-hospital disease outbreak.[35] At least in principle, such methods could be applied in the broader community outside hospitals as well.

## CHALLENGE 3: LACK OF EFFECTIVE PROPHYLAXIS AGAINST RESISTANT STRAINS

Under standard best practice, the drug isoniazid is given to family members of infected TB patients, to slow down the spread of the disease among those close to TB sufferers. While it protects many people from infection by isoniazid-susceptible strains of the disease, such prophylactic treatment does much less to protect people against isoniazid-resistant strains.[36] This gives isoniazid-resistant strains an advantage in the Transmission Game (outside hospitals), in addition to their advantage in the Treatment Game (inside hospitals). Worse still, the medical community has yet to identify any effective prophylactic treatment for those at risk of developing isoniazid-resistant disease.[37]

The lack of effective prophylaxis against highly resistant disease is a challenge for any "Detect + Search" strategy, since it makes halting transmission more difficult among those who are identified as being at risk of developing drug-resistant disease. Fortunately, transmissibility can be reduced in other ways as well. As Dr. Helen Cox of Médecins Sans Frontières explains: "A major part of it is education about TB transmission and cough hygiene—together with separate sleeping arrangements. The patient is also encouraged to wear a paper mask in overcrowded and closed conditions, and the caregiver is provided with N95 respirators."[38] Directing more resources to control transmission among those at risk of contracting drug-resistant TB could be an effective way to put resistant strains at a disadvantage in the context of a "Detect + Search" strategy, even without any antibiotic options for prophylaxis.

## CHALLENGE 4: THE FREE-RIDER PROBLEM

Unlike CRE, tuberculosis frequently transmits itself outside hospitals, circulating regionally and even globally. Consequently, no individual hospital can hope to make much of a dent, on its own, in the overall prevalence of XDR-TB. Moreover, even if hospitals are *collectively* capable of tackling XDR-TB, they may not have sufficient *individual* incentive to do so. Each hospital faces difficult decisions about how to allocate limited resources. Fielding the sort of transmission-fighting teams needed to implement an effective "Detect + Search" strategy against XDR-TB is costly.[39] An individual hospital that chose to bear these costs could undercut its own quality of care, while having only a minimal impact on the overall prevalence of XDR-TB.

In such a scenario, it's natural to expect hospitals to focus on their own patients first and let someone else worry about the overall resistance of disease. Indeed, we can think of hospitals as being in a Prisoners' Dilemma when it comes to combating XDR-TB, each with a dominant strategy not to field any transmission-fighting teams outside of their own facilities, but all worse off as XDR-TB is allowed to spread more freely.

The most natural solution to this Prisoners' Dilemma is "cartelization" (see chapter 3), under the leadership of authoritative third parties such as the CDC in the United States or the Ministry of Health and Family Welfare in India. (The World Health Organization could also play an important role by providing guidance on best practices to national health authorities, much as the CDC is providing guidance to hospitals in the fight against CRE.) These organizations are already experienced at controlling infectious diseases in the field and, given enough resources and public support, could organize, train, and deploy the army of transmission-fighters needed to win the war against rising antibiotic resistance.

# Let's Choose Victory

What if public health authorities don't act quickly or strongly enough, and the world is overrun with totally resistant bacteria? Thousands and perhaps even millions in the developed world could once again die from dreaded diseases like tuberculosis, while those in poorer places face a public health crisis even more dire than they do today. Even then, however, it won't be too late. As long as some susceptible bacteria remain in circulation, we can adopt a targeted transmission-fighting stance at any time, to push back against the prevalence of resistant disease. It will be a longer and tougher fight, but we can still win—eventually. Let's just hope and pray that it doesn't come to that. Let's nip total resistance in the bud, while it's still rare and at its most vulnerable. Let's take the relatively easy victory over antibiotic resistance while it's still within reach.

# The Triumph of Game-Awareness

The Game-Changer Files highlight the power of game theory, but it's worth noting how little formal analysis they contain. Each Game-Changer File is mostly devoted to understanding the games at hand as fully as possible. That's also how I spent the vast majority of my time preparing those Files, in a quest for greater game-awareness: poring over every information source I could imagine; studying the scientific literature, when relevant, and consulting experts to fill in the gaps; and even infiltrating online discussion boards (like the eBay Suspension Forum) to see how players in the game really think about it.

Without such preparation, I still could have written down a game-theory model and "solved" it. But my recommendations would not have been worth much. In the same way, when you apply what you have learned here, first take the time to "know what you don't know," then fill in those gaps as best you can. Finally, where possible, design your strategic plan to be robust and/or adaptable to any remaining uncertainties. Do that, and you'll enjoy the profound strategic advantage that game theory can provide.

If you devise your strategy based on an unrealistic simplification of the game, however, you could be in big trouble. Before the United States invaded Iraq in 2003, for instance, Deputy Defense Secretary Paul Wolfowitz testified before the House Budget Committee: "It's hard to conceive that it would take more forces to provide stability in post-Saddam Iraq than it would take to conduct the war itself and to secure the surrender of Saddam's security forces and his army. Hard to imagine." A better imagination might have helped our leaders understand the real game that we would face in post-Saddam Iraq and, with understanding, they might have been better prepared for the sort of counterinsurgency operation that would ultimately be needed to stabilize the country.

Fortunately, developing your game-awareness also naturally builds your imagination, so that you can avoid similar hazards in the games you play. By contrast, those who lack game-awareness remain prone to strategic snares, including those left by others to entrap them. This is the lesson of "Sun Bin's Revenge," a classic tale of woe to the game-unaware player (Pang Juan) who faces a more game-aware foe (Sun Bin).

## Example: Sun Bin's Revenge

> King Wei asked, "Is there a way to attack an enemy whose strength is ten times ours?" Master Sun Bin replied, "There is. Attack him where he is not prepared, and go by way of places where it would never occur to him you would go."
>
> —Sun Bin (died 316 BC), The Art of War

Sun Bin was a great military strategist of the Warring States Period in China, who lived about one hundred years after his more famous ancestor, Sun Tzu.[*] Sun Bin's longtime nemesis and foil was Pang

---

[*]Sun Tzu is arguably the greatest military strategist of all time, author of the classic *Art of War*, which is still taught in military colleges today. For many years, scholars suspected that Sun Tzu never existed and that Sun Bin wrote this treatise. However,

Juan, general of the Wei kingdom, the most powerful state at that time. They had begun as the closest of friends, "sworn brothers" and fellow students under the hermit Guigazi. But Pang betrayed Sun while both were serving as military advisers to King Hui of Wei, framing Sun for treason. As punishment, the king condemned Sun Bin to be tattooed across the face as a criminal, and to have his feet cut off. He would spend the rest of his life as a cripple.

Pang Juan's plan was to keep Sun Bin alive just long enough for Sun to compile his military knowledge into a book. Recognizing the need to escape, Sun feigned madness, even going so far as to gleefully devour animal feces when Pang had him locked in a pigsty to test his sanity. Failing to appreciate that Sun could be faking, Pang let down his guard, after which Sun managed to escape to the rival Qi kingdom, which he would serve for the rest of his career.

Years later, in 354 BC, Sun Bin got his first taste of revenge against Pang Juan. Pang had led Wei's powerful army to besiege the capital of the Zhao kingdom, who called upon the Qi kingdom for help. Rather than marching to defend Zhao, as Pang expected, Sun Bin took the fight straight to the Wei kingdom, attacking their capital. This forced Pang to abandon his siege on Zhao and rush back to defend his own king. Sun's forces ambushed them on the way, handing Pang a crushing defeat at the battle of Guilang.[1]

Sun Bin's victory at Guilang rested on Pang Juan's failure to imagine the possibility that Sun might choose *not* to defend his ally and instead attack Pang's capital. If Pang had anticipated this possibility, he could have easily neutralized it by leaving behind a small rear guard to protect the capital. (Such a contingent would only need to resist a siege by Sun's forces for a short period of time, just long enough for Pang to return and attack.) Having left no such guard behind, however, Pang was forced to scramble back home and, in his haste, was easy prey to Sun's ambush.

in 1972, researchers rediscovered Sun Bin's *The Art of War,* which differs from and builds upon Sun Tzu's (presumably earlier) work.

Sun Bin's final act of revenge was, if anything, sweeter than the first. Prior to yet another totally avoidable Wei defeat (at the battle of Maling), Sun Bin felled a tree across the road in an ambush area, carving the words "Pang Juan dies under this tree" on the trunk. Sun then positioned 10,000 of his finest crossbowmen on either side of the road, placing them on alert and saying, "After dark, when you see a torch lit, let your bolts fly." When Pang arrived later that night, he noticed the carvings and so lit a torch for a closer look. At the torch's lighting, the Qi crossbowmen launched a deadly volley that sent the Wei troops into panic, scattering them into retreat.

Realizing that he was out of options, Pang then cut his own throat, so the story is told, saying, "I may as well help the wretch make a name!" And, indeed, he did. Sun Bin is remembered to this day as being among the greatest strategists who ever lived—all because he always kept his eyes open to all of the possibilities.

## Open Your Eyes!

Blinding ignorance does mislead us. O! Wretched mortals,
open your eyes!

—*Leonardo da Vinci*

Like anything of great value, game-awareness is not easily achieved. It's a habit that must be constantly cultivated and maintained, and reading alone won't get you where you need to be. So, when you finish this book, practice. Practice until you see the game in everything, and everywhere, from the halls of Congress to the aisles of your local grocery store. Open your eyes!

# ACKNOWLEDGMENTS

I never could have written this book on my own.

For one thing, I first had to learn to open my own eyes and see the world of games for myself. For that, my gratitude goes to the friends who taught me what it means to be a game theorist. There are too many to list in full but allow me to honor five, in particular, whose generosity, kindness, and brilliance have long been an inspiration for me: Jeremy Bulow, my first graduate school mentor and also my boss at the Federal Trade Commission; Bob Wilson and Paul Milgrom, my dissertation co-advisers, each of whom I expect you will know as Nobel Laureates before too long; and Susan Athey and Bob Gibbons, two more potential Nobelists who counseled, advised, and encouraged me on many occasions.

The richness of this book is in its examples, many of which I learned about from others. My student Yacine Amrani told me how Tariq ibn Ziyad was the original boat-burner; the basic ideas for "The Cigarette Advertising Ban," "More Than Lizards," and "The Lockhorns' Night Out" are thanks to Professor Mikhael Shor (University of Connecticut, Economics Department), taken from lecture notes which he generously shared years ago; "College Admissions and the

College Board" grew out of discussions with Professor Chris Avery (Harvard Kennedy School of Government); Professor Leslie Marx (Duke Fuqua School of Business) told me the story of convict-turned-CEO Kuno Sommer, which appears in the endnotes to "Disrupting Harmful Cooperation"; "Addicts in the Emergency Department" builds on a final project for my game-theory class by the student team of Alex Kerr, Blake Lloyd, Dan Reese, and Sarah Schiavetti; and my student Takuya Sato told me about the battles of Guilang and Maling in "Sun Bin's Revenge."

For all my emphasis on the importance of humility when it comes to applying game theory in practice, this book is nothing if not audacious. I've dared to tackle real strategic problems in a wide variety of fields, including some about which I knew very little when I began. Fortunately, experts helped fill gaps in my understanding and ensure that my focus remained firmly on the "real games" at play. I cannot thank enough those whose input deepened my analysis of several applications, including some that didn't make it into the final draft but where I still want to acknowledge help that enriched my own understanding. (Of course, I take full credit for any remaining errors or omissions.)

- *"The END Amendment"*: Professor David Ridley (Duke Fuqua School of Business) told me about NanoViricides; Andrew Robertson and Rianna Stefanakis (BIO Ventures for Health) directed me to their work documenting the PRV pipeline.
- *"Dialing for Dollars"*: David Evans (*Bloomberg News*) gave feedback and posed questions that prompted a more careful analysis of the pros and cons of a fund-raising cartel.
- *"Real Estate Agency"*: Professors Steven Levitt (University of Chicago, Economics Department) and Chad Syverson (University of Chicago Booth School of Business) helped me to improve my discussion of their work on real estate, while Courtney James (Urban Durham Realty) offered an insightful real estate agent's perspective.

- *"Addicts in the Emergency Department"*: Professor Gail D'Onofrio (Yale School of Medicine, Emergency Medicine Department), Dr. Leonard Paulozzi (CDC), and Professor Anna Lembke (Stanford School of Medicine, Psychiatry Department) each shared invaluable facts and perspective on the practice of emergency medicine, as well as on the broader prescription drug epidemic. Professors Bill Boulding and Rick Staelin (Duke Fuqua School of Business) pointed me to their recent research on patient satisfaction, while Professor Seth Glickman (UNC School of Medicine, Emergency Medicine Department) shared early findings from Bivarus's pilot program at the UNC Hospitals.

- *"eBay Reputation"*: Professor Jonathan Levin (Stanford, Economics Department) pointed me to several eBay initiatives that share the spirit of my ideas, such as its Detailed Seller Ratings, while Google's Michael Schwarz told me about his own patent-pending online reputation system. Members of the eBay Research Labs also read an early version—the head of the group wrote me that "you caught our attention"—but were unable to provide any comments. Given the quality of the minds at work there—including Steven Tadelis, one of the best applied game theorists anywhere—I remain very curious to learn what they are developing to address the issues of seller fraud and buyer extortion.

- *"Antibiotic Resistance"*: Dr. Mario Raviglione, Director of the Stop TB Department of the World Health Organization, provided extensive help on the section dealing with Cepheid's Xpert system and corrected several misunderstandings I had about vaccines, tuberculosis transmission, and prophylaxis. Dr. Arjun Srinivasan, CDC associate director for Healthcare Associated Infection Prevention Programs, informed me of IFTAR's initiatives and highlighted the similarity between my ideas on targeted contact tracing and CDC's recent guidance on how to fight in-hospital CRE. Dr. Maria Joyce, an expert on medical microbiology and susceptibility testing at Duke Hospital, brought the

Xpert system to my attention. Kunal Rambhia, a senior analyst at the UPMC Center for Biosecurity, helped me appreciate the limits of vaccines and alerted me to the risk of conjugation. Attendees of the Duke Infectious Diseases Grand Rounds, the research seminar of the Duke Infectious Diseases Department, where this work was presented, also provided helpful comments.

- *"Urban Counter-Insurgency Warfare" (not in book)*: US Army Major Neil Hollenbeck has taught me much about counter-insurgency, from doctrine to operations at every level from rifle squads to divisions, and the attitudes and working relationships of American commanders with local security partners in Iraq and Afghanistan. (Neil also devised the "look" of the payoff matrices used throughout the book.)

- *"Dogs' Strategic Emergence" (not in book)*: The story of how the dog emerged from the wolf is a fascinating tale of strategic evolution or, I should say, coevolution of dog and man. Professors Bridgett vonHoldt (Princeton, Ecology and Evolutionary Biology Department), Robert Wayne (UCLA, Ecology and Evolutionary Biology Department), and Darcy Morey (Radford, Forensic Science Institute) helped me to understand recent genetic evidence coming out of India and China in light of earlier archeological findings and, more broadly, to appreciate the richness and subtlety of canid evolution.

- *"The Stability of Tit-for-Tat" (not in book)*: Professor Robert Axelrod (University of Michigan Ford School of Public Policy) directed me to findings on Tit-for-Tat's effectiveness when players' strategies are subject to random errors, an important consideration in strategic evolution.

These experts are leaders in their fields, people who you might not expect would answer unsolicited emails from a random economist. I was blown away by their generosity and willingness to help.

Of course, the most heartfelt thanks go to those closest to home. My wife Lesley held down the fort during the crazy months when this book was created—with a new baby, no less—and provided a sanity check against some of my wilder ideas. Thanks are also due to my editor Jack Repcheck, who gently guided me "away from the teacher/ student mind-set," to numerous friends and colleagues who provided comments on portions of the manuscript, and to the entire team at W. W. Norton who made this book possible.

Finally, I want to thank you, my readers, in advance, for encouraging me with stories of how you used game theory to transform your work, your home, your life for the better. (Please share these stories at McAdamsGameChanger.com.)

I wrote this book in the conviction that game theory can be a powerful force for good. Now, please prove me right.

Durham, North Carolina
June 2013

# NOTES

## Prologue

1. Raymond W. Smith, "Business as War Game: A Report from the Battle-front," *Fortune*, September 30, 1996.

## Introduction

1. Zhuge Liang is widely viewed as the greatest strategist of China's Three Kingdoms period, distinguishing himself as a brilliant scholar (writing military classics such as *Thirty-Six Stratagems* and *Mastering the Art of War*), inventor (credited with the world's first land mine *and* with *mantou*, a steamed bun still enjoyed today), military commander, and statesman.
2. Martin Kihn, "You Got Game Theory!," *Fast Company*, February 1, 2005.
3. See www.gametheory.net/links/consulting.html for a partial list of business strategy consulting firms that use game theory. (Unless mentioned otherwise, this and all other links in the notes were successfully accessed on April 30, 2013.)
4. "How Companies Respond to Competitors: A McKinsey Global Survey," *McKinsey Quarterly*, May 2008.
5. Tom Copeland is the author, with Vladimir Antikarov, of *Real Options: A Practitioner's Guide* (Cheshire, UK: Texere, 2001).
6. http://www.gallup.com/poll/1645/guns.aspx.
7. Spain granted this concession in the Treaty of Utrecht as a condition for

Britain joining the victorious alliance that opposed French–Spanish unification in the War of Spanish Succession.

8. See Peter Temin and Hans-Joachim Voth, "Riding the South Sea Bubble," *American Economic Review*, 2004, for a fascinating case study of one winner, C. Hoare and Co., a fledgling West End bank that made a profit of over £28,000.

9. For instance, traders who can forecast another trader's need to sell routinely engage in "predatory trading," forcing the one in need to accept greater losses than he would otherwise suffer. See Markus Brunnermeier and Lasse Pederson, "Predatory trading," *Journal of Finance*, 2005. Predatory trading may help explain why J. P. Morgan's $2 billion trading loss on credit default swaps in May 2012 was revised upward to $5.8 billion in July.

10. Miller received the Nobel Prize in 1990 for other pioneering work. Black died before receiving his prize.

11. This quote is taken from Roger Loewenstein, *When Genius Failed: The Rise and Fall of Long-Term Capital Management* (New York: Random House, 2000).

12. Ariel Rubinstein, "A Sceptic's Comment on the Study of Economics," *Economic Journal*, 2006.

## Chapter 1: Commit

1. "Gibraltar" is a Spanish derivation of Jabal Tariq ("The Mountain of Tariq").

2. Another reason Tariq probably didn't burn the fleet: it was the gift of an African ally. Burning the fleet would therefore have been an uncharacteristically impolitic move for Tariq, especially since he could have achieved the same effect by simply sending the ships back across the strait.

3. Hwajung Oh and Adrian Taylor, "Brisk Walking Reduces Ad Libitum Snacking in Regular Chocolate Eaters During a Workplace Simulation," *Appetite*, 2012.

4. Believe it or not, this business of having "two selves" is fairly standard economics. For decades, economists have grappled with the fact that people routinely make commitments whose only real effect is to restrict their own options. One of our leading theories now is the "dual self model." According to this model, motivated and supported by psychological and neurological evidence, each of us is really two selves: (i) an impulsive self who has default control over moment-by-moment actions and (ii) a cooler self who can step forward and take control on an as-needed basis. The impulsive self is easily swayed by temptation, while the cooler self is more able to resist. However, exercising control is tiring for the cooler self, so that it actually becomes harder to control oneself the longer that one must exercise control. This model helps to explain many diverse phenomena. For

example, why do alcoholics pour their drink down the drain? If they can muster the willpower to do that, surely they must have enough strength to resist taking a little sip? No—not if willpower is "like a muscle." When tired, in a moment of weakness, they will drink. Anticipating this, it makes sense for alcoholics to commit not to drink, or at least force themselves to take the extra step of going to buy more alcohol. See, e.g., Drew Fudenberg and David K. Levine, "Timing and Self-Control," *Econometrica*, 2012, presented as the Fisher–Schultz Lecture at the 2010 World Congress of the Econometric Society.

5. Ian Ayres and Barry Nalebuff, "Skin in the Game," *Forbes*, November 13, 2006.

6. "Microcomputers Catch on Fast," *BusinessWeek*, July 12, 1976.

7. This is a heavily condensed version of vixen25's actual statement, available at http://community.homeaway.com/thread/3381. As it turns out, vixen25's real problems began after the vacation ended, when the owner refused to return her $500 security deposit and played a dirty trick to get HomeAway to remove vixen25's complaint from their site. (According to vixen25, the owner offered to give back the deposit if vixen25 removed her complaint and, when she agreed, asked her to put that in writing. When vixen25 sent an email stating "I will remove my negative feedback once you return my deposit," the owner forwarded it to HomeAway, claiming that vixen25 was trying to extort her. HomeAway removed the complaint and the owner never returned the deposit.)

8. http://www.elliott.org/blog/vacation-rental-scams-are-a-growing-problem.

9. Airbnb's system also protects property owners from unscrupulous renters. First, Airbnb's reputation system allows property owners to offer feedback on renters, helping owners to identify (and not rent to) freeloaders who might try to stay one night and then make a false complaint. Second, Airbnb mediates all payments between owners and renters. This allows Airbnb to limit owners' flexibility to renegotiate the price. This stops unscrupulous renters from threatening payment cancellation or negative feedback to pressure the property owner to accept a lower price. Such "buyer extortion" is a problem on other sites, such as eBay, that bring buyers and sellers together but do not mediate the payment process. (For more on eBay, see Game-Changer File 5, "eBay Reputation.")

10. Geoffrey Fowler, "Airbnb Is Latest Start-Up to Secure $1 Billion Valuation," *Wall Street Journal*, July 26, 2011. By September 2012, Airbnb had a valuation of $2 billion; see Alyson Shontell, "Airbnb Raising $100 Million at a $2 Billion+ Valuation," SFGate.com, September 27, 2012.

11. Ty McMahan, "HomeAway's CEO Talks IPOs & Airbnb's Valuation," *Wall Street Journal*, October 13, 2011.

12. Videogame console makers incur huge R & D costs in developing new gaming systems. Once those costs are "sunk," console makers have an incentive to keep their systems on the market, even at prices too low for them to recover all their costs.

13. Nintendo also launched a me-too console in 2001, but the GameCube was a resounding failure, with only about 21 million units sold worldwide. In 2006, Nintendo changed course and carved out its own niche with the Wii, a console that doesn't even attempt to appeal to hardcore gamers. Despite conceding that market to the PS2 and Xbox, Nintendo's Wii led both Sony and Microsoft in cumulative sales as of the first quarter of 2012.

14. A player's payoff could incorporate many factors, including "social concerns" such as whether he/she did "better" than the other player, whether he/she and the other player acted "fairly," and so on. Players are *not* assumed to care only about themselves. Rather, the notion of "payoff" is defined to capture everything that matters to players, including fellow feeling.

15. The fact that both players get their third-best outcome instead of their second-best outcome is a general feature of all of the Prisoners' Dilemma games that we will see throughout the book.

16. This quote is a lightly edited combination of text from the abstract and introduction to Richard H. McAdams, "Beyond the Prisoners' Dilemma: Coordination, Game Theory, and Law," *Southern California Law Review*, 2009.

17. These escape routes from the Prisoners' Dilemma naturally combine and reinforce one another, e.g., the success of a cartel may hinge on its members' ability to retaliate, you and I may trust each other because we have a relationship and/or we may have a relationship because we trust each other, and so on. However, each escape route is conceptually distinct.

## Focus 1: The Timing of Moves

1. O. G. Haywood, Jr., "Military Decision and Game Theory," *Journal of Operations Research Society of America*, 1954. Figures 5 and 6 are taken from Haywood's article.

2. Kate Snow, "Obama, Clinton Ditch Press for Secret Meeting," *ABC News*, June 6, 2008.

3. As it turned out, McCain's team spent just a small fraction of that time vetting Sarah Palin, their actual VP pick. Senator McCain insists that more vetting would not have changed his decision but, nonetheless, the *perception* that McCain rushed his decision certainly didn't help his cause.

4. Signal jamming is the (usually deliberate) transmission of signals that disrupt communications by decreasing the signal-to-noise ratio. For instance, totalitarian regimes routinely censor foreign radio broadcasts by sending

out strong signals of their own over the same wavelengths, drowning out the undesired message in a great wash of noise.

## Chapter 2: Invite Regulation

1. The term "common resource" is a misnomer since the real issue in these problems is uncontrolled access, not whether a community owns the resource. Communities may sometimes be in the best position to exercise effective control over an exhaustible resource, as opposed to national or international regulation or a system of private property. See, e.g., Elinor Ostrom, *Governing the Commons: The Evolution of Institutions for Collective Action* (New York: Cambridge University Press, 1990).

2. Harvard appears to have played the most central role in the popularization of football in the United States. For one thing, Harvard was the first American college to adopt the rugby-influenced tackling rules that characterize the modern game, having learned that style of play during a cross-border match with McGill, a Canadian college. Harvard was also an enthusiastic early adopter of the sport, building a 30,000-plus-seat stadium (Soldiers Field) as early as 1904.

3. As a Harvard undergraduate, Roosevelt had a front-row seat during football's first years. Indeed, as a freshman, he witnessed the second ever Harvard–Yale football game. (Harvard lost.) See John J. Miller, "Teddy Roosevelt Becomes a Football Fan," *National Review Online*, April 12, 2011. For more on the early years of football and Roosevelt's role, see John J. Miller, *The Big Scrum: How Teddy Roosevelt Saved Football* (New York: HarperCollins, 2011).

4. The NCAA was founded to halt the worst violence on the football field, but now plays a much more expansive role. For a fascinating analysis of the economics of college sports and the NCAA's role, see Cecil Mackey, "College Sports," chapter 11 of Walter Adams and James Brock, eds., *The Structure of American Industry*, 9th edition (Englewood Cliffs, NJ: Prentice Hall, 1995).

5. US Senate, Consumer Subcommittee of the Committee on Commerce, "Cigarette Advertising and Labeling" hearing, 91st Congress, 1st Session, July 22, 1969.

6. These examples are taken from *Time*'s 2009 retrospective of famous cigarette ads: http://www.time.com/time/magazine/article/0,9171,1905530,00 .html.

7. In this, the advertising ban largely succeeded. The next major regulatory move against tobacco was in 1996, when the Food and Drug Administration acquired authority to regulate tobacco products.

8. Christopher Lydon, "Ban on TV Cigarette Ads Could Halt Free Spots Against Smoking," *New York Times*, August 16, 1970.

9. James L. Hamilton, "The Demand for Cigarettes: Advertising, the Health Scare, and the Cigarette Advertising Ban," *Review of Economics and Statistics*, 1972.

10. Since TV stations had to run two messages for every ad purchased by a cigarette company, they naturally charged cigarette companies more than other advertisers. In this way, the cigarette companies ultimately paid for the public-service announcements.

11. In the United States, praziquantel is most commonly prescribed to clear worms from dogs and cats, though at a hefty markup over its pennies-per-dose cost. For instance Bayer's Drontal Plus was available recently from 1-800-PET-MEDS for $14.99 per dose.

12. For details, see "Accelerating Work to Overcome the Global Impact of Neglected Tropical Diseases: A Roadmap for Implementation," World Health Organization, 2012.

13. "Merck Serono to Raise Donation Tenfold to Eliminate Worm Disease," Merck press release, January 30, 2012.

14. Michael Regnier, "Neglected Tropical Diseases: The London Declaration," Wellcome Trust blog, January 31, 2012, available at http://wellcome trust.wordpress.com/2012/01/31/neglected-tropical-diseases-the-london -declaration.

15. Priority review vouchers can be sold between firms, so one may expect them to be acquired by whatever firm has the biggest new drug farthest back in line for review. The original idea to award priority review vouchers was proposed by Duke University economists David Ridley, Henry Grabowski, and Jeffrey Moe in a 2006 article, "Developing Drugs for Developing Countries," published in *Health Affairs*. Senator Sam Brownback (R–KS) picked up the idea, shortly thereafter introducing bipartisan legislation with Senator Sherrod Brown (D–OH).

16. Moxidectin is currently used in animals to treat heartworm and intestinal worms, but has not yet been formulated for human use. For more on this and other promising developments, see Andrew S. Robertson, Rianna Stefanakis, Don Joseph, and Melinda Moree, "Analysis of Neglected Tropical Disease Drug and Vaccine Development Pipelines to Predict Issuance of FDA Priority Review Vouchers over the Next Decade," BIO Ventures for Global Health, 2012.

17. http://www.businesswire.com/news/home/20110222005851/en/Nano Viricides-Presented-anti-Dengue-Hemorrhagic-Fever-Studies-Dengue.

## Focus 2: Strategic Evolution

1. Both of Darwin's key theories—natural selection, which acts on individuals, and sexual selection, which acts on relationships—are relevant for understanding strategic evolution. My focus here is on natural selection. For a fascinating discussion of sexual selection, see Matt Ridley, *The Red Queen: Sex and the Evolution of Human Nature* (New York: HarperCollins, 1993).

2. Adam Davidson, "The Purpose of Spectacular Wealth, According to a Spectacularly Wealthy Guy," *New York Times*, May 1, 2012.

3. http://press.collegeboard.org/sat/faq, accessed December 19, 2012.

4. Jenny Anderson, "A Hamptons Summer: Beach, Horses and SAT Prep," *New York Times*, August 13, 2012.

5. In 2011, nineteen-year-old Sam Eshagoff took the SAT and/or ACT sixteen times for different students, charging about $2,500 per test. See "The Perfect Score: Cheating on the SAT," *60 Minutes*, January 1, 2012.

6. The National Center for Fair and Open Testing lists all four-year colleges and universities that are "ACT/SAT-optional," in the sense of not requiring ACT or SAT scores for admission from at least some students. See http://www.fairtest.org/university/optional. Among the top ten liberal arts colleges in *US News*'s 2012–13 ranking, only Bowdoin (ranked 6) is ACT/SAT-optional and, among the top fifty "national universities," only Wake Forest (ranked 27) and University of Texas, Austin (ranked 46) appeared on the list in September 2012. (Other high-ranking schools such as NYU (ranked 32) and Middlebury (ranked 4) also appear, but they only exempt students who can provide a substitute measure such as Advanced Placement tests.)

7. As one admissions dean explained, there are simpler, more effective, and perfectly legitimate ways to inflate reported SAT scores: "Go SAT-optional. Students who don't have to submit SAT scores only bother if the scores are high." See Daniel de Vise, "Claremont–McKenna SAT scandal: More at stake than rankings?" *Washington Post* blog, February 7, 2012.

8. Another concern is cheating by colleges themselves, several of which were recently caught falsifying their students' SAT scores. Baylor paid already-admitted students to retake the SAT (uncovered in 2008); Iona reported false acceptance rates, SAT scores, graduation rates, and alumni giving (2011); Claremont–McKenna inflated SAT scores and class rankings (2012); and Emory inflated SAT scores for more than a decade (2012). See Richard Perez-Pena and Daniel Slotnik, "Gaming the College Rankings," *New York Times*, January 31, 2012, and Laura Diamond and Craig Schneider, "Emory Scandal: Critics Doubt College Rankings," *Atlanta Journal-Constitution*, August 26, 2012. *US News* has recently announced that

Claremont–McKenna's number 9 ranking and Emory's number 20 ranking are unchanged, even after their falsified numbers are replaced with the true numbers. Indeed, Claremont–McKenna's lie amounted to about 10–20 points per test section, a relatively small effect.

9. William Lichten, "Whither Advanced Placement—Now?" in Philip M. Sadler et al., eds., *AP: A Critical Examination of the Advanced Placement Program* (Cambridge, MA: Harvard Education Press, 2010).

10. College Board, "Get with the Program," available at http://professionals.collegeboard.com/profdownload/ap-get-with-the-program-08.pdf.

11. Thomas Pfankuch, "Students Losing Full Advantage of Advanced Placement," *Times-Union*, June 23, 1997.

12. Ibid.

13. At Harvard in 2010–11, the following AP exams did not count toward Advanced Standing: Art (Studio and Portfolio), Comparative Government and Politics, Computer Science A, Environmental Science, Human Geography, and US Government and Politics. At Stanford in 2012–13, the list of credit-unworthy AP exams included all those (except Computer Science A) *plus* Biology, English Language and Composition, English Literature and Composition, European History, Italian Language and Culture, Macroeconomics, Microeconomics, Music Theory, Psychology, Spanish Literature and Culture, Statistics, US History, and World History. See "Advanced Standing at Harvard College: 2010–2011," Harvard University, and Stanford's "AP Credit Chart: 2012–2013," available at http://studentaffairs.stanford.edu/registrar/students/ap-charts.

14. "Average AP cut score [i.e., the minimal score for credit] has moved up about half a point between 1998 and 2006.... Acceptance scores of non-selective colleges have switched from 3 to a mixture of 3 and 4; selective colleges from a mixture of 3 and 4 to largely 4; and highly selective colleges from 4 to 4 and 5." Even as credit standards have gotten tougher, however, mean scores have fallen, from 3.10 in 1986 to 3.00 in 1996 and 2.89 in 2006. See William Lichten, "Equity and Excellence in the College Board Advanced Placement Program," *Teachers College Record*, 2007, and "AP Grade distributions—all subjects 1987–2006," College Board, 2006.

15. These quotes are from John Tierney, "AP Classes Are a Scam," *Atlantic*, October 13, 2012.

16. Another concern is that the proportion of prepared students who take AP courses is skewed to disadvantage some minorities. According to the College Board's "8th Annual AP Report to the Nation" (2012), only 20 percent of African American students it considers prepared took an AP course last year. That contrasts with 30 percent of prepared Hispanic students, 38 percent of prepared white students, and 58 percent of prepared Asian students.

17. Isaac Asimov, "My Own View," reprinted in *Asimov on Science Fiction* (Garden City, NY: Doubleday, 1981).

18. Barry Sinervo and C. M. Lively, "The Rock–Paper–Scissors Game and the Evolution of Alternative Male Strategies," *Nature*, March 21, 1996.

19. When blue-throats are common, the best "strategy" is to be an orange-throat, since orange-throats are stronger and can easily seize blue-throat territory. When most males are yellow-throats, however, being an orange-throat is the worst possible option since, with so many yellow-throats prowling around, you are likely to be cuckolded left and right. By contrast, blue-throats' monogamy protects them from cuckoldry.

## Chapter 3: Merge or "Collude"

1. In today's world of global business, mergers must also pass the muster of European regulators, who routinely take preemptive action against suspected cartel behavior. In 2012, European anti-cartel actions included several fines in the tens of millions of euros, including a €169-million fine on the freight-forwarding business, as well as unannounced inspections of firms producing everything from North Sea shrimp to plastic pipe fittings and optical disk drives. See http://ec.europa.eu/competition/cartels/cases/cases.html, accessed December 19, 2012.

2. See US Department of Justice and Federal Trade Commission, "Antitrust Guidelines for Collaboration Among Competitors," 2000, and "Horizontal Merger Guidelines," 2010.

3. See US Department of Justice and Federal Trade Commission, "Antitrust Guidelines for the Licensing of Intellectual Property," 1995. Some antitrust scholars have argued that these guidelines need to be updated in light of increasingly pervasive and complex IP licensing strategies. See, e.g., Joshua Newberg, "Antitrust, Patent Pools, and the Management of Uncertainty," available at http://www.ftc.gov/opp/intellect/020417joshuanewberg.pdf.

4. In *Federal Baseball Club v. National League* (1922), the Supreme Court—under Chief Justice William Howard Taft, a former college baseball player—ruled unanimously that major league baseball was exempt from the Sherman Antitrust Act. In *Radovich v. National Football League* (1957), however, the court ruled that football was not exempt. Congress responded in 1961 with the Professional Sports Broadcasting Act, allowing baseball, football, basketball, and hockey teams (but not boxing venues) to bargain collectively with radio and television broadcasters over broadcast rights.

5. College football's Bowl Championship Series (BCS) has been criticized on antitrust grounds, including by Senator Orrin Hatch (R–UT), who argued

on *Bloomberg TV* in March 2012 that the BCS's "privileged conferences" system constitutes an illegal restraint of trade. The mountain states have produced a string of strong teams with perfect seasons in recent years, such as Utah in 2009 and Boise State in 2010, who nonetheless did not get a chance to play in the BCS "national championship" game. For video of Hatch's remarks, see http://www.youtube.com/watch?v=ClZ7B2KfFtk. I discuss the farm cooperative exemption in "Dialing for Dollars."

6. In 1938, a typical diamond ring in the United States cost $80. The average American salary at the time (when only about half of American households had indoor plumbing) was $1,299.

7. For more on the Diamond Syndicate, see Edward Jay Epstein, "Have You Ever Tried to Sell a Diamond?," *Atlantic*, February 1, 1982.

8. Mr. Oppenheimer's full speech is published in an appendix to the business case: Debora Spar, "Forever: De Beers and US Antitrust Law," *Harvard Business School*, 2002.

9. De Beers has settled its regulatory problems in the United States by, among other things, pleading guilty in 2004 to price-fixing charges. This settlement enabled De Beers to sell diamonds directly in the United States. As of mid-2012, De Beers had ten retail locations in the US, from New York to Naples, Florida, and Costa Mesa, California.

10. Nicholas Stein, "The De Beers Story: A New Cut on an Old Monopoly," *Fortune*, February 19, 2001.

11. Even if diamonds lose their place as the symbol of choice for eternal love, they will remain valuable for adornment as well as for many industrial uses. Indeed, through its subsidiary Element Six, De Beers has become a world leader in the synthetic diamond market, squeezing out flawless lab-grown diamonds for a wide variety of industrial uses. As their website notes, "Synthetic diamond's combination of extreme properties such as thermal conductivity, chemical inertness and semi-conduction offer a huge range of opportunities," not to mention the extreme hardness that has long made diamonds ideal for cutting and crushing. For more on the industrial uses of synthetic diamonds, see http://www.e6.com/wps/wcm/connect/E6_Content _EN/Home. For a related business case on whether synthetic diamonds pose a threat to De Beers' core engagement-ring business, see David McAdams and Cate Reavis, "De Beers' Diamond Dilemma," MIT Sloan School of Management, Case 07-045, 2008.

12. David Evans, "Charities Deceive Donors Unaware Money Goes to a Telemarketer," *Bloomberg Markets*, September 12, 2012.

13. Donaldson also disputed the notion that the Cancer Society had lost money on its InfoCision campaign, telling *Bloomberg*: "I can state definitively that

the American Cancer Society's Notes to Neighbors program was indeed profitable for the society and never revenue negative."

14. Shortly after the InfoCision scandal broke in late 2012, InfoCision's soon-to-be-ex-CEO stated that "In better years, profits were about 10 percent of sales. Now, profits are down to 5 percent." See Betty Lin-Fisher, "Refocused InfoCision Tries to Move Forward," *Akron Beacon Journal*, November 10, 2012. While such numbers cannot be confirmed—as a privately held firm, InfoCision does not reveal its financial results—they appear plausible.

15. As Kelly Browning, CEO of the American Institute for Cancer Research, explained to *Bloomberg*: "They have the kind of scale required to do it and do it fairly efficiently."

16. Congress could potentially impose equally stringent disclosure rules on charities themselves, requiring them (say) to reveal the cost of every separate campaign rather than an averaged figure over all campaigns. However, it's natural to expect lawmakers and regulators to impose weaker rules on charities themselves than on for-profit middlemen. As explained in the text, requiring disclosure could effectively kill all of the more costly sorts of campaigns that a charity might benefit from conducting. Consequently, a legitimate case can be made that requiring disclosure could harm charities.

17. Many local United Way chapters named themselves "Community Chests," a creative marketing move that would be picked up by Monopoly, the popular board game. (According to Philip Orbanes's book *The Monopoly Companion*, the Community Chest in Atlantic City was located near Pacific Avenue, where you will find a Community Chest spot on the modern Monopoly board.)

18. It isn't obvious how antitrust authorities would view a joint venture among charities to merge and/or coordinate their telemarketing operations. Joint ventures motivated by efficiency considerations are generally evaluated less stringently than mergers, so it's possible that America Gives might not prompt an antitrust challenge. However, as discussed in the text, such a combination could drive for-profit telemarketers out of the market, foreclosing competition and harming charities left out of the cartel.

19. With the growth of competition in transportation, the original 1920s justification for the farmer cooperative exemption no longer applies. This has led regulators and leaders in Congress to speak openly about revoking farmer cooperatives' antitrust exemption. For instance, in Senate Judiciary Committee testimony in 2009, Attorney General Christine Varney stated to chairman Patrick Leahy: "Well, Senator, it does seem to me that an examination of whether the law is serving its intended purposes may lead to a conclusion that it is not the right law for the state of the industry at this time," to

which Senator Leahy (D–VT) replied, "I've [been] appointed as chairman of the Antitrust Subcommittee in the Judiciary Committee. [Eliminating the farm cooperative exemption] is something that I know [Senator Kohl (D–WI)] and I will be discussing, and with Senator Sanders (I–VT)." See "Crisis on the Farm: The State of Competition and Prospects for Sustainability in the Northeast Dairy Industry," hearing before the Committee on the Judiciary, United States Senate, September 19, 2009.

20. The biggest charities in 2011 were Lutheran Services of America ($18.3 billion in revenue), the Mayo Clinic ($8.0 billion), and the YMCA ($5.9 billion). See "The 200 Largest US Charities," *Forbes*, available at http://www.forbes.com/lists/2011/14/200-largest-us-charities-11_rank.html.

21. Since the business of America Gives is making phone calls, an extra wrinkle that might arise relates to freedom of speech. In recent years, InfoCision's client list has included the National Republican Senatorial Committee (NRSC) who, *Bloomberg* reports, paid InfoCision $115 million from 2003 to 2012. Suppose that, some political season, the Republican Party were (say) to push to eliminate the charitable tax deduction. Would it be constitutional for Congress to pass a law that compels America Gives to place fund-raising calls on behalf of the NRSC, against its own economic interest? Another reason not to preclude all discrimination is the potential efficiency gain from employing more highly skilled solicitors who are especially knowledgeable and motivated to raise funds for specific fields, such as women's health, medical research, homelessness, etc. Fund-raising cartels could better utilize such skilled and motivated solicitors by restricting membership to charities within the same field. However, this is only possible if they "discriminate" against all other charities.

## Focus 3: Equilibrium Concepts

1. In that case, there would also be a third Nash equilibrium in which Leroy and Loretta each adopt a "mixed strategy," i.e., in which they each *randomly* decide what to do. For an excellent discussion of mixed strategies, a topic beyond the scope of this book, see chapter 7 on "unpredictability" in Avinash Dixit and Barry Nalebuff, *Thinking Strategically* (New York: Norton, 1991).

2. This notion that monopolists can earn more by selling less will be familiar to anyone who has studied microeconomics, but it may seem strange to others. The basic idea is that monopolists profit from creating scarcity in their own products. That way, as people clamor for what they're selling, the Monopolist can charge a higher price.

3. Reinhard Selten won the 1994 Nobel Prize in Economics (alongside John

Nash and fellow equilibrium-theory pioneer John Harsanyi) for identifying an equilibrium concept for sequential-move games. I will refer to Selten's concept as "Rollback equilibrium," after the method one uses to find such equilibria. (Game theorists often use a different name: "Subgame-perfect equilibrium.")

4. There is no generally accepted name for this equilibrium concept. Some game theorists use the phrase "strategic moves." Others correctly note that what I'm calling a "Commitment equilibrium" is just the Rollback equilibrium of an enriched game in which one player moves first (committing itself) *and* moves last (implementing that commitment) in a three-stage game.

5. These ideas were formally developed in a revolutionary set of papers by Stanford's "Gang of Four"—David Kreps, Paul Milgrom, John Roberts, and Robert Wilson—published together in 1982 in the *Journal of Economic Theory*: Paul Milgrom and John Roberts, "Predation, Reputation, and Entry Deterrence"; David Kreps and Robert Wilson, "Reputation and Imperfect Information"; and "Rational Cooperation in the Finitely Repeated Prisoners' Dilemma" by the whole gang. These are among the most important papers in economic theory not yet recognized with a Nobel Prize.

## Chapter 4: Enable Retaliation

1. Obviously, this tale takes quite a few liberties with the history of the Wild West. Jesse James never met Billy the Kid and, even if he had, they would have had little interest in killing one another. Indeed, by most accounts, Billy the Kid was actually a pretty nice guy. The Long Branch Saloon was real, famous as the site of many shootouts, gunfights, and standoffs.

2. The term "Mexican Standoff" refers to a gunfight in which the first person to shoot is sure to die. Perhaps the most famous example is the dramatic final scene in the classic spaghetti western *The Good, the Bad, and the Ugly*. In this scene, the title characters are locked in a three-way duel. Whoever shoots first can only kill one adversary, leaving the third man time to gun the first shooter down. Consequently, no one wants to draw first, leaving director Sergio Leone plenty of time to build up to perhaps the most dramatic (anti)climax in cinematic history. For the movie lovers out there, I can't resist adding that this scene is even more interesting, from a game-theory point of view. Previously, without Tuco's knowledge, Blondie had emptied the bullets from Tuco's gun. So, what appeared to be a three-way standoff between Blondie, Tuco, and Angel Eyes was actually a two-way duel between Blondie and Angel Eyes, only with Angel Eyes at a disadvantage since he *thought* Tuco was also in the game.

3. This fictional atomic bomb was so eerily similar to the real bomb under

development in America's super-secret Manhattan Project that the FBI immediately launched an investigation. When Campbell and Cartmill were interrogated, the FBI was suspicious of their claims to have independently hit upon the Manhattan Project design. It didn't help that Cartmill lived in *Manhattan* Beach, California, a connection that the FBI's lead agent found far too curious to be coincidental. For details, see Robert Silverberg, "Reflections: The Cleve Cartmill Affair: One," *Asimov's Science Fiction*, March 2010, available at http://www.asimovs.com/_issue_0310/ref.shtml.

4. In his autobiography *I. Asimov: A Memoir"* (1995), legendary science fiction author Isaac Asimov refers to Campbell as "the most powerful force in science fiction ever, and for the first ten years of his editorship he dominated the field completely." Campbell also wrote science fiction, including one classic tale of an Antarctic science mission gone bad that has been made into a movie three times, most recently as *The Thing* (2011).

5. The first detonation of a nuclear bomb, codenamed "Trinity," occurred over the White Sands Proving Grounds in New Mexico.

6. Natural Resources Defense Council, "Table of Global Nuclear Weapons Stockpiles, 1945–2002," available at http://www.nrdc.org/nuclear/nudb/datab19.asp, accessed December 19, 2012.

7. Karl W. Uchrinscko, "Threat and Opportunity: The Soviet View of the Strategic Defense Initiative," Department of the Navy, Naval Postgraduate School, PhD thesis, 1986, available at http://www.dtic.mil/cgi-bin/GetTRDoc?Location=U2&doc=GetTRDoc.pdf&AD=ADA178649.

8. The imaginary story told here is at least as far-fetched as some of the Hollywood movies mentioned earlier. My point is just to illustrate the possibility that unsteady resolve to retaliate could prompt a nuclear war, even if neither side really wants to attack the other. (In the story, President #1 could be mistaken and the Soviets might not want to attack after all. However, the Soviets still attack because they believe that President #1 believes that they will attack.) The MAD danger ratchets up considerably should one or both sides truly desire to strike the other.

9. $100 is the so-called "marginal cost" of serving an additional passenger. Other "fixed costs" of operating on the Chicago–Atlanta route are not accounted for here. As long as the number of flights on the route is fixed, these other costs are irrelevant to the firm's pricing decision.

10. Since the airlines compete repeatedly for travelers, they may still be able to escape this Prisoners' Dilemma by leveraging their "relationship" to keep prices high. See chapter 6.

11. By a similar logic, prices may also rise together. An airline that raises its price and is not immediately matched has a natural incentive to lower

its price back to the original level. Anticipating this, airlines may exploit dynamically posted prices to follow one another up or down the price ladder.

## Chapter 5: Build Trust

1. http://www.snopes.com/fraud/advancefee/nigeria.asp.
2. As Stephen Diamond, a clinical and forensic psychologist, explained in 2009 in *Psychology Today*: "The term 'psychopath' was replaced at some point in psychiatry by 'sociopath,' in part to try to lessen its social stigma. . . . Any time you hear the terms psychopath, sociopath, asocial, amoral or dissocial personality, the appropriate corresponding diagnosis [as specified in the Diagnostic and Statistical Manual of Mental Disorders of the American Psychiatric Association] may (or may not) be Antisocial Personality Disorder." See Stephen Diamond, "Masks of Sanity (Part Four): What is a Psychopath?," *Psychology Today* blog, August 31, 2009.
3. For more, see Martha Stout, *The Sociopath Next Door* (New York: Broadway Books, 2005), as well as Heather Clitheroe's entertaining review: http://www.bookslut.com/scarlet_woman_of_selfhelp/2005_03_004676.php.
4. Seena Fazel and John Danesh, "Serious Mental Disorder in 23,000 Prisoners: A Systematic Review of 62 Surveys," *Lancet*, 2002. Also intriguing, though more speculative, is the notion that sociopaths are more prevalent in the executive suite than in the general population. That said, recent media claims that "10% of Wall Street Employees are Psychopaths" are junk. See Dr. John Grohol, "Untrue: 1 out of Every 10 Wall Street Employees is a Psychopath," *Psych Central*, March 6, 2012, available at http://psychcentral.com/blog/archives/2012/03/06/untrue-1-out -of-every-10-wall-street-employees-is-a-psychopath.
5. The price of narcotics varies wildly over time and from location to location and, in any event, is difficult to observe directly. One recent estimate put the wholesale price of one kilo of cocaine in New York City at $23,000. See http://www.narcoticnews.com/Cocaine-Prices-in-the-U.S.A.php, accessed December 19, 2012.
6. David Gluckman, "A Guide to Certified Used Car Programs: Someone Was Kind Enough to Pre-Own Your Next New Car," *Car and Driver*, March 2009.
7. "Aston Martin Company History: 1930–1939," available at www.aston martin.com/the-company/history.
8. A few were priced so ridiculously low (e.g., a 2000 Mercedes-Benz S430 with 102,460 miles, originally priced at $75,000, sold for $4,250 on eBay) that one suspects damage or some other serious defect might have played a role in these particular examples.

9. See "Aston Martin Launches 'Assured' Seal of Approval," *Aston Martin News*, July 10, 2009.

10. It would be interesting to compare a 2001 Vantage V12 coupe sold through the Assured program with the one that *Popular Mechanics* saw sell on eBay. However, none of the Assured vehicles that I found were built prior to 2007.

11. Jim Mateja, "Which is the Better CPO Buy: Luxury or Economy?" available at http://www.cars.com/go/advice/shopping/cpo/stories/story.jsp?story=lux Econ.

12. Jim Mateja, "Is CPO for You?," available at http://www.cars.com/go/advice/shopping/cpo/stories/story.jsp?story=cpoForYou.

13. There is also a matter of trust. Since third parties who offer extended warranties aren't concerned about your repeat business, they may throw up roadblocks to make it difficult for you to collect when you need to invoke your warranty. Consequently, a dealer or factory warranty may be superior to one bought on the open market. Also, since dealers and/or the manufacturer themselves pay if there is a problem with a CPO vehicle, they have an incentive to screen which cars are certified, often with access to their complete maintenance history. Without access to such rich data on past performance, independent mechanics can't do nearly as good a job at assessing a vehicle's true quality.

## Chapter 6: Leverage Relationships

1. Henry Schneider, "Agency Problems and Reputation in Expert Services: Evidence from Auto Repair," *Journal of Industrial Economics*, 2012.

2. Scott D. Hammond, Deputy Assistant Attorney General for Criminal Enforcement, Antitrust Division, US Department of Justice, "The Evolution of Criminal Antitrust Enforcement over the Last Two Decades," speech at the 24th Annual National Institute on White Collar Crime, February 2010, available at http://www.justice.gov/atr/public/speeches/255515.pdf.

3. DOJ also offers an Individual Leniency Program that guarantees complete amnesty to any individual who confesses, even if his/her firm does not. The availability of individual leniency puts even greater pressure on firms to confess once *anyone* in their organization becomes aware of illegal activity.

4. That said, even the threat of jail time sometimes isn't enough to ensure cooperation in an antitrust investigation. In 1997, the Swiss health care giant Hoffmann–La Roche (HLR) was caught in a cartel fixing the price of citric acid. They quickly pleaded guilty, paid a $14 million fine, and promised to cooperate. However, when the DOJ deposed Kuno Sommer, HLR's director of worldwide marketing for vitamins, he lied under oath about the existence of a much larger conspiracy to fix the price of vitamins. Dr. Som-

mer likely felt that there was little reason to fear being caught in this lie, as the vitamin cartel (referred to by cartel members as "Vitamins Inc.") was extraordinarily disciplined in its secrecy. Indeed, according to Gary Spratling, head of antitrust criminal enforcement at the DOJ, top executives "took great pains to conceal [cartel] activity," issuing instructions to destroy all records and notes of their meetings or face possible termination. Two years later, however, co-conspirator Rhône-Poulenc confessed to its participation in Vitamins Inc., HLR paid a $500 million fine (the biggest ever), and Kuno Sommer was sentenced to four months in US prison for perjury. (See Stephen Labaton and David Barboza, "US Outlines How Makers of Vitamins Fixed Global Prices," *New York Times*, May 21, 1999.) Interestingly, being a convicted criminal has been little more than a speed bump in Dr. Sommer's career. Shortly after leaving prison, he was named CEO of Berna Biotech, a global vaccine company, and beginning in April 2012, he served as chairman of the Bachem Group, a "biochemicals company providing full service to the pharma and biotech industry."

5. It's hard to imagine that J. Edgar Hoover's FBI was entirely unaware of the Mafia. Why didn't they investigate more deeply and raise the alarm years earlier? Hoover's onetime number three man, William C. Sullivan, explains it this way in *The Bureau: My Thirty Years in Hoover's FBI* (New York: Norton, 1979): "The Mafia is . . . so powerful that entire police forces or even a mayor's office can be under Mafia control. That's why Hoover was afraid to let us tackle it. He was afraid that we'd show up poorly."

6. We can only speculate what those "continuing benefits" might have been. The continued safety and prosperity of one's family is an obvious possibility. But perhaps the most important factor, at least for some mafiosi, was personal honor itself. The older generation of gangsters stuck especially fiercely to traditions that they felt were "right" even when it cost them, refusing to sell drugs (even after younger mobsters proved the profitability of that trade) and insisting that their men grow no facial hair (for which they suffered derision from the younger generation, who mocked them as "Mustache Petes").

7. Known as "the Dapper Don" for his fancy suits and quick wit with reporters, Gotti routinely hosted Mafia get-togethers at the Ravenite Social Club in Little Italy, where they could be easily seen by reporters and arrested en masse by law enforcement.

8. See Philip Carlo, *Gaspipe: Confessions of a Mafia Boss* (New York: HarperCollins, 2009).

9. See O. Gefen and N. Q. Balaban, "The Importance of Being Persistent: Heterogeneity of Bacterial Populations Under Antibiotic Stress," *FEMS Microbiology Review*, 2009.

10. The cheating mechanism within biofilms is complex and nuanced. For more details, see S. Diggle, A. Griffin, G. Campbell, and S. West, "Cooperation and Conflict in Quorum-Sensing Bacterial Populations," *Nature*, November 15, 2007, and K. Sandoz, S. Mitzimberg, and M. Schuster, "Social Cheating in Pseudomonas Aeruginosa Quorum Sensing," *Proceedings of the National Academy of Sciences*, 2007.

11. Through a process known as lateral gene transfer, bacteria within the same film share genes with one another. Consequently, individual bacteria benefit when the film as a whole thrives, as well as from their own individual success. Since "cheater genes" imperil the entire film, this could help explain why genetic mutations that induce cheating behavior may not enjoy a long-run reproductive advantage, even if they have a short-run advantage during the life of a single film.

12. In Axelrod's tournament, each match lasted 200 rounds. Having a known end date complicates the game in ways that are not addressed in my brief summary here. For our purposes, think of each match as always being likely (though not certain) to last several more rounds.

13. In Axelrod's game, one round of punishment is sufficient. In other games, where the gain from cheating is larger or the cost of being punished is less, the Tit-for-Tat strategy will punish cheaters for more than one round before offering forgiveness.

14. For details, see Robert Axelrod, *The Evolution of Cooperation*, revised edition (New York: Basic Books, 2006). Since Axelrod's seminal analysis, game theorists have shown that TFT's success hinges on the assumption that players never misunderstand other players' moves. "When mistakes in communication are possible, Tit for Tat breaks down. Once a mistake in communication leads one side to believe that the other has defected, both sides will get into a mindless and almost slapstick routine of punishing each other: player A punishes B for defecting and then B punishes A for punishing him and so on [in a never-ending cycle]. This suggests that the desirability of Tit for Tat may not be robust to situations where players' actions may be misinterpreted." See Barry Nalebuff, "Puzzles: Noisy Prisoners, Manhattan Locations, and More," *Journal of Economic Perspectives*, 1987.

15. TFT may never catch on with isolated players on the fringes of the social network, who are not well connected to its tightly knit center. To the extent that cooperative success is necessary to acquire the social capital to forge even more links, such fringe players can get stuck in a vicious cycle, whereby their isolation causes their relationships to fail and their relationship failures cause them to remain isolated.

## Summary: How to Escape the Prisoners' Dilemma

1. In a more realistic setting, the prisoners and/or their associates will likely interact again, either inside or outside of prison. If so, the game would have repeated moves, in the sense that the prisoners can potentially link what happens in this game to what will happen in future games.

## File 1: Price Comparison Sites

1. Among large retailers, Costco insulates itself from the threat of showrooming by forming partnerships with suppliers to offer uniquely packaged versions of popular products. Similarly, small boutiques still thrive by offering unique items that are difficult to find online.

2. See Brad Tuttle, "Is Amazon Due for a Backlash Because of Its 'Evil' Price Check App?" *Time: Moneyland*, December 13, 2011.

3. Jeffrey Brown and Austan Goolsbee, "Does the Internet Make Markets More Competitive? Evidence from the Life Insurance Industry," *Journal of Political Economy*, 2002.

4. ATPCO's only "competitor" is SITA, which distributes some fares in Africa, Asia, and Europe.

5. "Competitive Impact Statement," *USA v. ATPCO et al.*, March 1994, 13–15.

6. For an in-depth analysis of Internet price obfuscation, see Glenn Ellison and Sara Ellison, "Search, Obfuscation, and Price Elasticities on the Internet," *Econometrica*, 2009.

7. BPPhoto.com's bricks-and-mortar location appears to be in Brooklyn, long a hotbed for the "gray market" in electronic equipment. (See http://donwiss.com/pictures/BrooklynStores/h0008.htm.) According to the *New York Times*, "Complaints [about Brooklyn electronics sellers] submitted to investigative agencies describe tactics like promising low prices but canceling orders or making threats when customers decline to add batteries and other accessories to their purchases." Added Anthony Barbera, manager of the Information and Investigations Department of the Better Business Bureau of New York: "It is a perennial problem for us, particularly in New York, not necessarily for New York customers but for customers around the country," made worse by the fact that these firms routinely shed their names and change their online identities. See Michael Brick, "In a Flash, Camera Dealers Feel the Web's Wrath," *New York Times*, January 11, 2006. That said, I want to emphasize that I am unaware of any wrongdoing by BPPhoto itself. Indeed, many customers have submitted positive feedback about BPPhoto ... along with a few who complain about tactics such as being forced to buy accessories. (For instance, snowdog_1 complained on resellerratings.com on January 21,

2013: "Very disappointed . . . the salesperson 'convinced' me to purchase two extended life batteries and two high capacity memory cards.")

8. Interestingly, $386.43 was exactly the same as the lowest price offered by any "used & refurbished seller," suggesting that BPPhoto knew exactly how to work the system and get listed as having the lowest overall price.

9. If PriceGrabber were a web-scraper (that finds and links to prices on the Web), its business model would require moment-by-moment price changes, since external websites can change on a moment's notice. However, PriceGrabber and sites like it actually get their prices through relationships with sellers, and could easily set the terms by which sellers can update posted prices.

10. Some restrictions can be justified on efficiency grounds. For instance, Price-Grabber users need to have confidence that the prices displayed on the site are accurate and available. Consequently, PriceGrabber is perfectly justified in kicking out retailers who fail to meet its quality standards, or excluding those who lack the capability to meet those standards.

## File 2: Cod Collapse

1. For more on early cod fishing in Newfoundland, see Heather Pringle, "Cabot, Cod, and the Colonists," *Canadian Geographic*, July/August 1997.

2. Kenneth Frank et al., "Trophic Cascades in a Formerly Cod-Dominated Ecosystem," *Science*, June 10, 2005.

3. As it turns out, herring and other small-fish cod prey (like capelin) are not really "winners" here. With the cod gone, nonfish competitors such as crab and shrimp now dominate the environment.

4. If a fisherman catches more valuable fish near the end of a trip, he can increase profit by keeping those fish and discarding less valuable (dead) fish that were caught previously. The law obligates him instead to throw back the valuable fish, leaving it alive in the sea. For evidence on discarding, see Ransom Myers et al., "Why Do Fish Stocks Collapse? The Example of Cod in Atlantic Canada," *Ecological Applications*, 1997.

5. Without sonar, fishing becomes futile once a population dwindles sufficiently. Fishermen then move on to another target, allowing the diminished stock to bounce back. With sonar, fishermen can identify exactly where to cast, making it profitable to hunt down even the very last fish.

## File 3: Real Estate Agency

1. Steven Levitt and Chad Syverson, "Market Distortions When Agents Are Better Informed: The Value of Information in Real Estate Transactions," *Review of Economics and Statistics,* 2008.

2. The investment properties owned by real estate agents likely differed from the homes of their clients, but Levitt and Syverson controlled for such differences to make an "apples-to-apples" comparison.

3. Another possibility is that homeowners may simply dislike living in a "show-ready" home. In my own case, for instance, the thought of prepping our kid-filled house for frequent showings was so overwhelming that we rented another place in which to live while our home was on the market.

4. Homes on city blocks with more uniform sales prices sold at a "discount" of 2.9 percent relative to what real estate agents got for their own similar properties (and took 2.5 fewer days to sell), while those on city blocks with more varied prices sold at a discount of 4.9 percent (and took two weeks less to sell).

5. Most agency contracts last for three to six months, after which the seller may switch agents. Recognizing this, agents know that they need to help close a deal relatively quickly, or risk losing everything.

6. Are agents who encourage their clients to sell quickly violating their fiduciary duty "to act at all times solely in the best interests of the principal"? Levitt and Syverson certainly think so, casting agents as "distorting information to mislead clients," but I suspect that most real estate agents see things differently. The real estate agents that I know make a genuine effort to provide excellent service, often going above and beyond to ensure that their clients are satisfied with the home sale process. Yet even these agents, who simply aim to please, might also prefer for their clients to set their asking prices a bit below the market. Why? Homeowners can't easily judge whether they got a good price for their home, but they can certainly tell if their home is taking longer than others to sell. Consequently, homeowners will tend to be more satisfied with their agents when their homes sell more quickly. Setting a below-market asking price can therefore be an effective way for agents to satisfy their clients, since doing so draws in buyers for a quick sale. A low starting price may also "anchor" homeowners' expectations about what price they are likely to receive, allowing them to feel satisfied even if the final sale price turns out to be low. Fundamentally, the problem here for homeowners is that what "satisfies" them (selling their home quickly) may not be good for them. Game-Changer File 4, "Addicts in the Emergency Department," explores a similar problem that arises in emergency departments, where hospitals' drive to "satisfy" patients may in some cases actually decrease the quality of care.

7. Agencies could hire specialists who are expert in managing these services, freeing up agents for what they do best.

8. Could a homeowner "piggyback" on his agent's relationship with contractors to get a good price? Perhaps not. The agent can introduce the homeowner to a high-quality contractor, for which the contractor has a strong

incentive to reward the agent, e.g., by providing low-cost services at *her* house. (Offering the homeowner a deal would undoubtedly also please the agent, but not so much as a deal for herself.) In fact, to the extent that the agent's recommendation signals quality, distinguishing a contractor from the field, that contractor knows that he can likely get away with charging an even higher price than usual. So, you might actually wind up paying *more* than if you'd found the same contractor's name by just flipping through the phone book.

9.  Real estate agencies are a perennial target of antitrust regulators. For example, in 2005, the Justice Department launched an investigation into "possible anticompetitive conduct in the provision of real estate services in the Tulsa [Oklahoma] area." This investigation centered on full-commission agents' practice of "boycotting," wherein they steer buyers away from homes offered by discount brokers. One such discount broker, J. D. Smith, abandoned his low-price policy and went back to charging the customary "3 + 3" commission structure. As he told *Money*: "In one week, I've had more showings and more offers from other realtors than I had in the previous two months." See Jon Birger, "Feds Probe Real Estate Agents," *Money*, April 22, 2005.

10. These quotes were taken from www.remax.com, when I accessed the site on May 24, 2012.

11. One such agency where I live in Durham, NC, is Urban Durham (urbandurhamrealty.com), founded by an ambitious and hardworking agent who jumped ship several years ago from one of the bigger agencies. Urban Durham signs are now everywhere in my neighborhood, where just a few years ago there were none.

12. In July 2012, NAR reported total membership of 993,715.

13. When buyers' agents are paid by the seller, all agents are identical in the buyer's eyes with regard to the fees that must be paid. No matter what home he winds up purchasing, the buyer will pay the same fee regardless of who represents him. Buyers then naturally choose to work with whomever they meet first. Consequently, the *real* business of real estate agency has much to do with building social connections, in order to be the person who gets introduced first to a new buyer. This makes it difficult to break into the high end of the real estate business, especially for those who lack a prestigious social pedigree.

14. The term "pay-to-play" (or "pay-for-play") comes from the radio industry, after record labels' longtime practice of paying DJs to play their songs. Such practices are now illegal. For an excellent economic analysis of the radio industry, see Peter Alexander, "Music Recording," in Walter Adams and James Brock, eds., *The Structure of American Industry*, 11th edition (Upper Saddle River, NJ: Pearson Prentice Hall, 2004).

## File 4: Addicts in the Emergency Department

1. Catherine Saint Louis, "ER Doctors Face Quandary on Painkillers," *New York Times*, April 30, 2012.

2. Charles Fooe, MD, Bat Masterson, and Marian Wilson, "ED Pain Management Program Hinders Drug-Seeking," *Emergency Medicine News*, January 2011.

3. Anna Lembke, "Why Doctors Prescribe Opioids to Known Opioid Abusers," *New England Journal of Medicine*, 2012.

4. Institute of Medicine, "Hospital-Based Emergency Care: At the Breaking Point," Washington, DC: National Academy of Sciences, 2006. This report contains an estimate that, across the nation, an ambulance is diverted due to overcrowding about once a minute, i.e., over half a million times per year.

5. Witness President Barack Obama, who exhausted much of his 2008–12 term of office passing the Affordable Care Act, which subsidizes an expansion of health insurance coverage and creates health insurance exchanges to spur competition, and his Republican opposition, who spent much of their own oxygen denouncing it.

6. See http://www.cdc.gov/media/pressrel/2010/r100617.htm. As Dr. Leonard Paulozzi of the CDC clarified to me, these numbers "are not a tally of 'drug seekers.' [The rising number of] visits for nonmedical use of prescription pain relievers [is] an indication that opioid abuse has grown, but we don't really know the trend in 'drug-seeking' in EDs," personal correspondence, August 19, 2012.

7. This scenario may sound contrived, but emergency physicians routinely see patients who claim allergies to all non-addictive painkillers.

8. Catherine Saint Louis, "ER Doctors Face Quandary on Painkillers," *New York Times*, April 30, 2012. All other quotes from Professor Gail D'Onofrio, Dr. Abhi Mehrothan, and Dr. Benzoni come from this source.

9. As of August 2012, forty-one states have operational PDMPs and forty-nine states (all but Missouri) have passed laws authorizing one.

10. Many PDMP databases require a special web portal and can be time-consuming to access. Fortunately, improvements are on the way that will allow emergency physicians to check prescription status more quickly and easily, undoubtedly inducing more doctors to check their PDMP and deny more drug-seekers. My point is that PDMPs do not *entirely* solve the problem, as emergency physicians still have a perverse incentive not to get tough on drug-seekers.

11. Even PDMPs don't remove all uncertainty, as some patients may indeed have an untreated chronic condition that brings them repeatedly to the emergency department in real pain.

12. Press release available at http://www.majorhospital.org/newsite/Infodesk/ AboutUs/topPressGaneyRankings.pdf.

13. http://www.epmonthly.com/whitecoat/2009/12/could-satisfaction-surveys -be-harming-patient-care/. In this same survey, "Eighty-one percent of medical providers were aware of instances in which patients intentionally provided inaccurate derogatory information on a satisfaction survey and 84% felt that patients used the threat of negative satisfaction surveys to obtain inappropriate medical care."

14. Some experts on patient satisfaction might disagree. See Joshua Fenton et al., "The Cost of Satisfaction: A National Study of Patient Satisfaction, Health Care Utilization, Expenditures, and Mortality," *Archives of Internal Medicine*, 2012, which finds that greater satisfaction is correlated with worse health outcomes, including an elevated risk of death, especially among those who self-report as being in excellent health. More recently, other researchers have argued that these findings are unreliable and that higher satisfaction—at least on some dimensions, such as satisfaction with nursing-staff communication— tends to be positively correlated with health outcomes. See Matthew Manary et al., "The Patient Experience and Health Outcomes," *New England Journal of Medicine*, 2013.

15. Adam Kilgore, "Stephen Strasburg Goes Deep," *Washington Post*, March 16, 2013.

16. This may be especially true of Stephen Strasburg. In 2010, *Sports Illustrated* described Strasburg as the "most hyped and closely watched pitching prospect in the history of baseball." (Tom Verducci, "Nationals Taking Safe Road with Strasburg but is it Right One?" *Sports Illustrated*, May 18, 2010.) However, perhaps in part due to what some analysts describe as "disastrous" pitching mechanics (see Lindsay Berra, "Throwdown: A Comparison of Stephen Strasburg and Greg Maddux's Pitching Mechanics," *ESPN the Magazine*, March 23, 2012), Strasburg was seriously injured during his first season in the majors. Strasburg returned successfully to the mound in 2012, when he was selected to play in the 2012 All-Star Game, despite (perhaps because of) the Nationals' decision to limit him to only about five innings per game.

17. Another potential concern is that doctors who care about their satisfaction scores may hesitate before raising uncomfortable but medically necessary topics of conversation. For example, suppose that an obese patient comes to the emergency department, suffering from an acute asthma attack. It might help this patient to know that obesity has been associated with asthma, but broaching such a subject might upset the patient and lead to poor satisfaction scores.

18. William Sullivan and Joe DeLucia, "2 + 2 = 7? Seven Things You May Not

Know About Press Ganey Statistics," *Emergency Physicians Monthly,* September 2010.

19. Worse still for hospitals, their own emergency physicians have ample financial incentive to blow the whistle on such practices. According to Sullivan and DeLucia, "Health care providers who are able to prove how pressures to improve patient satisfaction scores unjustifiably increased costs to Medicare and Medicaid may choose to file 'whistleblower' lawsuits in hopes of earning up to 30% of the recovered overpayments hospitals receive. Any perceived retaliation against providers who file [such] lawsuits subjects hospitals to even further liability."

20. In place of Press Ganey, hospitals could and should shift their attention to other measures of patient satisfaction (and other forms of patient feedback) with more proven links to clinical outcomes. For instance, one recent study found that patient satisfaction with the clarity of their discharge instructions is associated with a decreased risk of hospital readmission within thirty days. See William Boulding et al., "Relationship Between Patient Satisfaction and Hospital Readmission Within 30 Days," *American Journal of Managed Care,* 2011. Rewarding discharge staff when patients are more satisfied with their discharge instructions could therefore make good sense, as a means to keep patients healthy and out of the hospital.

21. Of course, Press Ganey surveys could still be useful in informing and motivating emergency physicians, but this needs to be done with an eye to serving patients, not pandering to them. Indeed, doctors ought to be rewarded for receiving negative Press Ganey scores, as long as those bad reviews are earned for the right reasons. For instance, we should applaud emergency physicians who receive complaints—such as "I expected a narcotic painkiller but didn't get one" or "I wanted a month's worth of drugs but only got four days"—that reflect a willingness to push back against the epidemic of prescription drug abuse.

22. "About Bivarus, Inc," available at http://www.linkedin.com/company/bivarus-inc-.

23. How do they do it? First, whereas Press Ganey's survey arrives in the mail a few days after discharge, Bivarus's survey is immediately accessible via smartphone or over the Web. (One might suspect that using the smartphone would discourage low-income patients from responding. However, according to Seth Glickman, assistant professor of Emergency Medicine at UNC and Bivarus cofounder, "Our data with smartphones actually suggests the opposite: it's an increasingly effective way to reach traditionally under-served patient populations [including] non-English speakers and racial minorities." Personal correspondence, March 2013.) Even better, the

Bivarus survey can be customized to the patient's experience, making the feedback process more meaningful for both doctors and patients.

24. The way in which health care is *received* by patients can be just as important as how it is *delivered* by providers. For instance, for type-2 diabetes sufferers, achieving good glycemic control requires diligent "self-monitoring of blood glucose, dietary restrictions, regular foot care and ophthalmic examinations," as well as adherence to prescribed medications, only possible when diabetes patients understand and *own* their own care. Unfortunately, according to the WHO's CODE-2 (Cost of Diabetes in Europe—type 2) study, only 28 percent of European patients treated for diabetes achieved good glycemic control while, in the US, less than 2 percent of adults with diabetes perform the full level of care that has been recommended by the American Diabetes Association. Worse yet, "poor adherence to recognized standards of care is the principal cause of development of complications of diabetes and their associated individual, societal and economic costs." (Quotes are from the World Health Organization's 2003 report, "Adherence to Long-Term Therapies: Evidence for Action.")

25. According to the Drug Abuse Warning Network (DAWN)'s "2010 Findings on Drug-Related Emergency Department Visits," there were 4.9 million drug-related ED visits in 2010 (including illicit drugs and alcohol). Over one-third of these were due to "misuse or abuse of pharmaceuticals." See http://www.samhsa.gov/data/2k12/DAWN096/SR096EDHighlights2010 .htm.

26. For more on SBIRT, see http://medicine.yale.edu/emergencymed/research/ sbirtalcohol.aspx.

## File 5: eBay Reputation

1. Another possibility is that criminals could be using these phony gift-card transactions to launder money.

2. The fourth, sold at a Buy It Now price of $97.95 with free shipping, might have been real or might have been the cleverest phony of all. After all, setting your Buy It Now price below $100 offers perfect cover, while setting it close to $100 (and above all other real prices) essentially guarantees that no real buyer will ever purchase.

3. Crooks don't need to possess a gift card to steal its value. As scambusters .org explains, at many stores they just need to obtain the card's unique serial number: "Imagine that a scammer comes into a store that displays gift cards on public racks (such as Wal-Mart) with a small and inexpensive mag-strip scanner in his pocket. This scanner can easily read and store the [unique serial number on each card]. Next, real customers come in to buy

some of these gift cards. Every few days, [the crook] calls the gift card phone number and enters the unique numbers to find out which cards have been activated, and what the remaining balances are.... The scammer can then go on a shopping spree and drain the gift card balances." See http://www.scambusters.org/giftcard.html. Responding to concerns about such gift card theft, Senator Chuck Schumer wrote a letter to the National Retail Federation and the Retail Gift Card Association in December 2011, stating: "Retailers who sell gift cards should take any and all security measures necessary to prevent the general public from stealing gift card numbers." See Whit Richardson, "Retailers Asked to Tackle Gift Card Fraud," *Security Director News,* December 20, 2011. Retailers have responded by modifying gift card packaging, making it harder for would-be thieves to read the serial numbers. Yet even that may not be enough. According to Adrian Pastor, principal security consultant with Corsaire, a boutique security consultancy, at least two UK retail chains were vulnerable to a "brute force" attack in which Corsaire was able to *guess* the serial numbers and access the value stored on those chains' gift cards, without ever needing to gain physical access to the cards. See Kelly Jackson Higgins, "Gift Cards Convenient and Easy to Hack," *Dark Reading,* October 23, 2009.

4. See "Holiday Gift Card Sales Reach All-Time High," NRF Press Release, November 2006. With iTunes gift cards, there are extra risks. In 2009, Apple began cracking down on iTunes gift cards that had been exchanged in violation of its terms and conditions. (See Ginny Mies, "Apple Cracks Down on Gift Card Fraud," *PCWorld,* June 23, 2009.) Those who use such gift cards may even have their iTunes accounts permanently disabled.

5. This quote is an edited and condensed version of an eBay guide written by julian640_0. The original version is at http://reviews.ebay.ie/Can-You-Still-Trust-eBay-Feedback?ugid=10000000004744070. eBay guides are written by eBay users and hosted on eBay's site.

6. John Gantz et al., "The Risks of Obtaining and Using Pirated Software," IDC White Paper (sponsored by Microsoft), October 2006, available at http://www.microsoft.com/en-us/download/details.aspx?id=3981.

7. Agam Shah, "Fake Apple iPad Knockoffs Sold on eBay for £50," *TechWorld,* August 9, 2010.

8. In October 2004, the scam was reported in Great Britain: John Leyden, "eBay 'Second Chance' Fraud Reaches UK," *The Register,* October 5, 2004. An August 2011 eBay Motors example is described in a video at http://www.youtube.com/watch?v=BOGHTRBHUAw.

9. Larry Barrett, "Another eBay Pirate Heads to Prison," InternetNews.com, January 14, 2010; and Mary Flood, "Prison for Houston Man Who Ran eBay Hot Tub Scam," *Houston Chronicle,* May 19, 2010.

10. http://www.strat-talk.com/forum/stratocaster-discussion-forum/107962-ebay-scammer-again.html.

11. http://www.aspkin.com/forums/ebay-accounts-sale/50423-selling-verified-usa-ebay-paypal-seller-accounts-increased-limits-100-5000-a.html, accessed September 28, 2012. How do the cheaters who frequent this site avoid being cheated themselves? The forum is curated by "aspkin" and "Greenbean," whom sellers thank profusely at the beginning of most posts for allowing them to offer eBay usernames through the site. Anyone who cheated a buyer would most likely be banned—and banned for real—losing the opportunity for future profits.

12. Having the ability to mutually retract negative feedback allowed buyers and sellers to remove the stain of negative feedback from their public profiles after resolving their dispute. When eBay removed sellers' ability to leave negative feedback, they also removed this mutual withdrawal option. Quoting from eBay's May 2008 announcement of the changes: "After much consideration, we've made the decision to remove the Mutual Feedback Withdrawal process. The reason is that—under the new rules—it opens sellers up to extortion." See "A Message from Brian Burke—Upcoming Feedback Changes," eBay announcement on May 7, 2008, available at http://www2.ebay.com/aw/core/200805191013132.html. That said, eBay rules still allow buyers to unilaterally modify negative feedback. Thus, buyer-extorters can still potentially demand unfair price reductions in exchange for modifying their feedback. See eBay rules on "Revising a Seller's Feedback" at http://pages.ebay.com/help/feedback/revise-feedback.html.

13. Anne P. Mitchell, "eBay to Stop Sellers from Posting Negative Buyer Ratings," *Internet Patrol*, June 19, 2009, available at http://www.theinternetpatrol.com/ebay-to-stop-sellers-from-posting-negative-buyer-ratings.

14. Michael Schwarz is a researcher in Google's "Strategic Technologies" group, as well as my friend and research coauthor. His idea is described in "Establishing and Updating Reputation Scores in Online Participatory Systems," Patent Application 20090006115.

15. The idea of weighting feedback by transaction volume was (to the best of my knowledge) first proposed in Jennifer Brown and John Morgan, "Reputation in Online Auctions: The Market for Trust," *California Management Review*, 2006.

16. Assuming that the seller never gets negative feedback and doesn't ever prepay more than the initial $100, the whole bank will be depleted once the seller has generated $100 worth of commissions. At that point, he will then pay all further commissions on a per-transaction basis and, as more commissions are paid, his reputation score will rise above 100.

17. Scores will continue to convey information about the *volume* of transactions

that sellers conduct on eBay. It is natural to suspect that most high-volume sellers are trustworthy, but we have seen examples (e.g., BruntDog's guitar dealer) of high-volume sellers who might still be worth avoiding.

18. To address this concern, at least partially, one could raise the threshold for what counts as negative feedback when it comes to determining sellers' reputation scores. For example, eBay might require proof that a fraud has occurred. However, doing so would create its own problems, as (i) crooks switch to schemes that cheat buyers in difficult-to-prove ways and (ii) unscrupulous buyers still attempt to extort honest sellers with the threat of fraud charges.

19. The reason that buyers pay two-way shipping costs is to ensure that honest sellers don't incur a loss should buyers return their merchandise due to "buyer's remorse."

20. A case might be made to mandate Extra Assurance in selected markets. The reason is that, as long as Extra Assurance is not widely adopted, there will be room for crooks to "blend in" and continue to defraud eBay buyers. If so, requiring all sellers to adopt Extra Assurance might be the only way to rid some eBay markets of seller fraud.

21. Law enforcement officers could catch such fraud by documenting some of the valuable stamps in the lot and then checking whether those stamps are included if/when the lot is returned.

22. Glen Stephens, "eBay Can Be a Great Stamp Resource," *Stamp News Australasia*, June 2011, available at http://www.glenstephens.com/snjune11.html.

23. eBay Item 190781531187, described at http://www.ebay.com/itm/Antique-Belgian-8-Day-Grandfather-Clock-Circa-1787-/190781531187?pt=LH_DefaultDomain_0&hash=item2c6b772433, accessed January 10, 2013.

24. The exact rule that determines when a transaction is "closed" is not essential.

25. eBay already offers buyers the option to submit anonymous feedback in the form of "detailed seller ratings" which are averaged to form aggregate measures of several aspects of the sales experience, such as accuracy of product description, shipping speed, and so on.

26. Unscrupulous buyers could still slam sellers in their response to feedback. However, as long as negative *responses* don't count as negative *feedback* when it comes to the seller's eBay reputation, sellers won't have much reason to worry about any such retaliation in the responses.

27. On the other hand, we don't want reporting extortion claims to be so easy that unscrupulous sellers submit false reports. To see the point, imagine what might happen if sellers could "anonymously" (and privately) alert eBay to buyer extortion, and imagine further that an unscrupulous seller has cheated a buyer and fears that the buyer will offer truthful feedback. Since there is always a chance that the buyer won't bother to complain, such

a seller has an incentive not to pick a fight by making a (false) public claim of buyer extortion. On the other hand, the seller does have an incentive to submit a false extortion alert to eBay, since doing so is costless and could provide "insurance" should the buyer complain that he has been scammed. (If the buyer truthfully claims fraud and the seller falsely claims extortion, their dispute will devolve into a "he said, she said" argument that eBay may not be able to resolve. In this way, claiming that the other side has cheated you can potentially serve as "insurance" if you have actually cheated them.)

## File 6: Antibiotic Resistance

1. Allison Aiello et al., "Consumer Antibacterial Soaps: Effective or Just Risky?" *Clinical Infectious Diseases*, 2007.

2. See Katie Paul et al., "Short-term Exposure to Triclosan Decreases Thyroxine *In Vivo* via Upregulation of Hepatic Catabolism in Young Long-Evans Rats," *Toxicology Science*, 2010, for one study on rats.

3. http://www.cdc.gov/biomonitoring/Triclosan_FactSheet.html.

4. http://www.fda.gov/downloads/ForConsumers/ConsumerUpdates/UCM206222.pdf. The FDA's hesitancy in banning triclosan appears to be due in part to the fact that triclosan can be effective at killing bacteria in some products. For instance, it is used in toothpaste to kill gingivitis.

5. In whack-a-mole, a classic carnival game, "moles" pop out of holes in a board and the player tries to whack as many as possible with a mallet. The frustrating—and farcical—element of the game is that the moles move just fast enough that they often disappear right as you are about to whack them.

6. http://www.mayoclinic.com/health/hand-washing/HQ00407.

7. "2011 Summary Report on Antimicrobials Sold or Distributed for Use in Food-Producing Animals," Food and Drug Administration and Department of Health and Human Services.

8. Jim Avila, "Superbug Dangers in Chicken Linked to 8 Million At-Risk Women," *ABC News*, July 11, 2012.

9. This process, known as "conjugation," allows diseases in the same host to share their drug resistance. For instance, drug resistance in a disease that only afflicts chickens could spread (in chickens) to one that makes both chickens and people mildly sick and then (in people) to a deadly and highly virulent human disease.

10. The controversy over triclosan began in 2007 and reached a climax in November 2010, when House Rules Committee chairwoman Louise Slaughter wrote a public letter to the FDA arguing that "triclosan should be banned in all consumer and personal care products." See http://www.louise.house .gov/images/stories/FDA_Letter_re_Triclosan_11-16-10.pdf. In July 2011,

Bath & Body Works was specifically targeted (by the activist group Beyond Pesticides) for its decision to launch a new line of antibacterial liquid hand soaps. See http://www.beyondpesticides.org/dailynewsblog/?p=5671. As of December 2012, Bath & Body Works still had not responded and, more to the point, to the best of my knowledge still did not offer any triclosan-free alternative in its stores or on its website (apart from specialty soaps such as an "aromatherapy" line).

11. See "Johnson & Johnson: Our Safety and Care Commitment" at http://www.safetyandcarecommitment.com/ingredient-info/other/triclosan.

12. This and other signs of vampirism, such as long fingernails, are actually normal by-products of bodily decay after death. But that didn't stop people from cutting out and burning the hearts of the newly deceased. See Paul Sledzik and Nicholas Bellantoni, "Bioarcheological and Biocultural Evidence for the New England Vampire Folk Belief," *American Journal of Physical Anthropology*, 1994.

13. Katherine Rowland, "Totally Drug-Resistant TB Emerges in India," *Nature*, January 13, 2012.

14. Resistance needs to be "stabilized" since, when it first arises, the changes that impart resistance often put bacteria at a disadvantage in other ways. Dr. Andersson's point here is that, over time, those disadvantages tend to evolve away. Consequently, if resistance is not countered quickly enough, it may never go away, even if the antibiotics that caused it to arise are no longer prescribed.

15. See Dan Andersson, "The Biological Cost of Mutational Antibiotic Resistance: Any Practical Conclusions?," *Current Opinion in Microbiology*, 2006. For an example of stable resistance in practice, see M. Sundqvist et al., "Little Evidence for Reversibility of Trimethoprim Resistance After a Drastic Reduction in Trimethoprim Use," *Journal of Antimicrobial Chemotherapy*, 2010.

16. See Ed Yong, "Fighting Evolution with Evolution—Using Viruses to Target Drug-Resistant Bacteria," *Discover*, May 2011.

17. Vaccine development is Goal 11.2 of IFTAR's "Public Health Action Plan to Combat Antimicrobial Resistance: 2012 Update," available at http://www.cdc.gov/drugresistance/actionplan/taskforce.html.

18. Even better, from the perspective of battling resistance, would be a vaccine that targets only the resistant strains of a disease. To the best of my knowledge, however, no such resistance-targeted vaccine currently exists. Furthermore, there are significant technical hurdles to overcome.

19. Any approach that limits drug-resistant bacteria's exposure to drug treatment can reverse unstable resistance. In addition to vaccines, other approaches include: (i) prescribing less unnecessary medication, as encour-

aged by the CDC's "Get Smart: Know When Antibiotics Work" program; (ii) diagnosing the drug susceptibility of a disease before prescribing treatment (more on this later); (iii) restricting the use of a specific drug to which bacteria have begun to develop resistance. An example of this last approach: CDC issued new guidance in August 2012 for the treatment of gonorrhea, specifying that doctors should no longer prescribe cefixime, the traditional first-line treatment. As Dr. Kevin Fenton, director of the CDC's National Center for HIV/AIDS, Viral Hepatitis, STD and TB Prevention, explained, "As cefixime is losing its effectiveness as a treatment for gonorrhea infections, this change is a critical preemptive strike to preserve ceftriaxone, our last proven treatment option.... Changing how we treat infections now may buy the time needed to develop new treatment options." See "CDC No Longer Recommends Oral Drug for Gonorrhea Treatment: Change Is Critical to Preserve Last Effective Treatment Option," CDC press release, August 9, 2012.

20.  This cycle of rising drug resistance is not a foregone conclusion, even if doctors don't change their prescribing habits. In particular, as long as the resistant strains that emerge are less infective and/or less transmissible than susceptible strains, we can expect resistant disease to remain relatively rare. This is what has happened (so far) with tuberculosis. Despite sixty years of antibiotic use against TB, the vast majority of cases remain susceptible to antibiotic treatment.

21.  "Cepheid Receives FDA Clearance for Xpert Flu," April 26, 2011, available at http://www.infectioncontroltoday.com/news/2011/04/cepheid-receives -fda-clearance-for-xpert-flu.aspx.

22.  Xpert MTB/RIF detects tuberculosis much more effectively than the old "smear microscopy" technique, which required visual detection of the bacterium under a microscope. According to the USAID press release: "Smear microscopy is particularly insensitive for diagnosing TB in patients who are co-infected with HIV." This has been a serious limitation since TB–HIV co-infections are widespread and, indeed, TB is the leading cause of death among people living with HIV in Africa.

23.  Like many antibiotics, rifampin is derived from molecules produced by bacteria themselves, weapons evolved over billions of years of microbial battle. (In rifampin's case, our benefactor was a bacterium found in soil taken from the French Riviera in the 1950s.) Unfortunately, this also means that disease-causing bacteria have already long faced and fought back against rifampin, developing defenses that can be tapped to resist its antibiotic effect. This helps explain why rifampin resistance tends to develop quickly in monotherapy (single-drug) treatment and why rifampin is often prescribed as part of a broader cocktail of drugs. See James Long, "Essential Guide to Prescription Drugs: 1992," pp. 925–29.

24. As of January 2013, Xpert MTB/RIF still had not been approved for use in the United States.

25. See "Public–Private Partnership Announces Immediate 40 Percent Cost Reduction for Rapid TB Test," USAID press release, August 6, 2012, available at http://www.usaid.gov/news-information/press-releases/public-private -partnership-announces-immediate-40-percent-cost.

26. There are several reasons why removing susceptible strains from the bacterial population may advantage the remaining resistant strains, causing them to grow in number more quickly than otherwise. For instance, suppose that prior infection by a susceptible strain prepares the immune system to fight all subsequent infections more successfully. Once susceptible strains disappear, the remaining resistant strains will be able to defeat the immune system more easily. Researchers have also identified a "crowding effect" when susceptible and resistant strains coexist within the same host, that removing the susceptible strains can free the remaining resistant strains to grow in number. (This phenomenon is known as "competitive release.") See, e.g., Andrew Wargo et al., "Competitive Release and Facilitation of Drug-Resistant Parasites After Therapeutic Chemotherapy in a Rodent Malaria Model," *Proceedings of the National Academy of Sciences*, 2007.

27. When dealing with highly resistant disease, there may not be an effective prophylactic treatment. In that case, isolation may be necessary to slow down transmission.

28. Since full participation is not essential, the proposal here is fundamentally different from steps routinely taken to combat outbreaks of highly infectious diseases. In those cases, it is essential to *contain* the disease. By contrast, when the goal is merely to disadvantage resistant strains, rather than stop all disease, containment is unnecessary. This is important, since containment requires extreme steps like quarantine that could create resentment and undermine the political viability of the program.

29. Mitchell Schwaber, Boaz Lev, Avi Israeli, et al., "Containment of a Country-Wide Outbreak of Carbapenem-Resistant Klebsiella Pneumonia in Israeli Hospitals via a Nationally Implemented Intervention," *Clinical Infectious Diseases*, 2011.

30. See "Guidance for Control of Carbapenem-Resistant Enterobacteriaceae (CRE): 2012 CRE Toolkit," CDC Division of Healthcare Quality Promotion, June 2012.

31. It takes two to three days to test a patient for CRE. For hospitals facing a greater incidence of CRE, the CDC suggests preemptively isolating all new at-risk patients until test results show they have no CRE.

32. "Clinical microbiology laboratories have often found it difficult to achieve accurate susceptibility testing results for carbapenem drugs." See Fred

Tenover et al., "Carbapenem Resistance in Klebsiella Pneumoniae Not Detected by Automated Susceptibility Testing," *Emerging Infectious Diseases*, 2006.

33. See Neel Gandhi et al., "Extensively Drug-Resistant Tuberculosis as a Cause of Death in Patients Co-Infected with Tuberculosis and HIV in a Rural Area of South Africa," *Lancet*, 2006.

34. In this case, the XDR-TB strains in question were resistant to ethambutol, streptomycin, aminoglycosides, and fluoroquinolones, in addition to isoniazid and rifampin.

35. See Mark Walker and Scott Beatson, "Outsmarting Outbreaks," *Science*, November 30, 2012.

36. See R. D. Fairshter et al., "Failure of Isoniazid Prophylaxis After Exposure to Isoniazid-Resistant Tuberculosis," *American Review of Respiratory Disease*, 1975.

37. Attempts to use rifampin or other drugs for prophylaxis have failed on a number of occasions. See John Livengood et al., "Isoniazid-Resistant Tuberculosis," *JAMA*, 1985, for a classic study along these lines, and S. H. Lee, et al., "Adverse Events and Development of Tuberculosis After 4 months of Rifampicin Prophylaxis in a Tuberculosis Outbreak," *Epidemiology and Infection*, 2012, for a more recent example.

38. See Theo Smart, "Managing MDR-TB in the Community: From Presentation to Cure or End-of-Life Care," *NAM Aidsmap*, October 18, 2010, available at http://www.aidsmap.com/Managing-MDR-TB-in-the-community -from-presentation-to-cure-or-end-of-life-care/page/1523027.

39. That said, hospitals have plenty of incentive to field *inward-facing* teams to eradicate XDR-TB within their own facilities. If enough TB spreads within hospitals, such within-hospital efforts could collectively have a meaningful impact on the wider spread of XDR-TB.

## Epilogue

1. Sun Bin's victory at Guilang was immortalized in the classic military treatise *Thirty-Six Stratagems* by Zhuge Liang. The phrase "besiege Wei to rescue Zhao" is still used in China today, much as Americans say "the best defense is a good offense."

# INDEX